Hands-on Signal Analysis with Python

Thomas Haslwanter

# Hands-on Signal Analysis with Python

## An Introduction

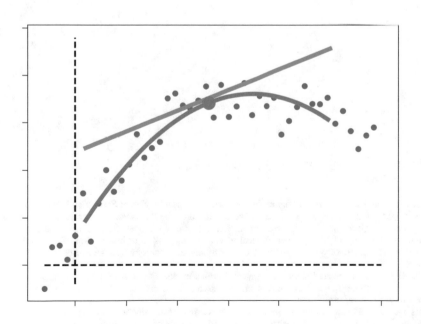

Springer

Thomas Haslwanter
Leonding, Austria

ISBN 978-3-030-57905-0          ISBN 978-3-030-57903-6   (eBook)
https://doi.org/10.1007/978-3-030-57903-6

This Springer imprint is published by the registered company Springer Nature Switzerland AG
The registered company address is: Gewerbestrasse 11, 6330 Cham, Switzerland

# Dedication

*For my wife Jean*
*Simply the best!*

# Preface

*"Data is the new oil"*, said the mathematician Clive Humby, who with his wife made \$140'000'000 helping UK retailer Tesco with its Clubcard system. Just as crude oil is not very useful when it comes out of the ground, it can be refined into amazing things, such as fertilizer, gasoline, plastic, etc. when processed correctly. The same is true for many signals: in their unprocessed form, they don't really help much. But when we extract the correct parameters, we can detect diseases, measure brain waves, predict the stock market or future health problems, etc.

This book aims to provide a good starting point for practical signal processing with Python. The programming language Python has become very popular, and a lot of information is available to get started with the language. For example, `http://scipy-lectures.org/` is a fantastic free starting point for scientific applications of Python. And additional assistance for the scientific use of Python can be obtained from good introductory books such as [Scopatz & Huff, 2015]. Excellent technical literature is also available on digital signal processing, such as [Smith, 1997], [Lyons, 2011], or [Rangayyan, 2002]. However, it is much harder to find literature combining the two fields. While [Unpingco, 2014] does so, he focuses on more technical aspects of the Fourier transform and Finite Impulse Response (FIR) filters.

## For Whom This Book Is

This book is for people with some previous programming experience, who now —for their graduate degree or for their research—want to use Python for signal processing. It tries to bridge the fields of programming and signal processing, and to provide a toolbox that allows the reader to solve most practical tasks that come up in practice. For this, it includes all the programs and examples used in this book, as well as the solutions to the exercises. It is by no means a comprehensive coverage of digital signal processing. By trying to provide a practical introduction to a wide range of topics, the book by default has to avoid detailed detours into the mathematical background.

A rudimentary technical background is assumed. For example, no introduction is given to basic linear algebra or on how to formulate for-loops, and technical information is provided only to the point that the reader knows what is happening, and that she knows where to look for technical details.[1] Those

---

[1] From here on I will refer to the reader as "her".

details are not presented here, but are left to authors more competent than me, such as [Smith, 2007a].

Many of the examples are taken from the field of life sciences, but the same principles are valid for other applications. In many cases, the goal of the desired data analysis can be formulated simply as follows:

- "I want to find repetitive patterns in my signals."
- "I want to construct a GUI (graphical user interface) for inspecting data."
- "I want to eliminate 50/60 Hz noise out from my measurement signals."
- "I want find the heart-beat variability from an ECG (electro-cardiogram)."

This book intends to provide the tools to achieve such goals quickly. Specifically, it will show how to

- Read in and write out data
- Display and visually inspect the data
- Smooth, filter, and integrate data
- Find events and the rate of those events
- Find patterns in data
- Compare groups to each other, or to a given value
- Fit lines, polynomials, and curves to data
- Evaluate the quality of those fits
- Interpret Fourier coefficients, and find the power spectrum of your signals
- Solve equations of motion by using the Laplace transform

To achieve all those goals as quickly as possible, the last chapter of the book provides hints on how to efficiently develop correct and working code. This should get you to the point where you can get things done quickly. A very brief outlook is also given to the popular topic "machine learning", so that the interested reader knows how to start, and where to look for further information.

## Acknowledgements

I wanted to thank Chris Bockisch, whose careful reading and helpful comments on the first few chapters really improved my writing style; Robert Merwa, whose thorough understanding of the Fourier Transform made that section more accurate and much clearer; and my wife Jean, who helped me with debugging, with the chapter on Statistics, and with her experience in GUIs.

Leonding, Austria                                                                        Thomas Haslwanter

## References

Lyons, R. G. (2011). *Understanding Digital Signal Processing*. Pearson.

Rangayyan, R. M. (2002). *Biomedical Signal Analysis*. Piscataway, NJ: IEEE Press.

Scopatz, A. & Huff, K. (2015). *Effective computation in physics*. O'Reilly Media.

Smith, III, J. O. (2007a). *Introduction to Digital Filters: with Audio Applications*. W3K Publishing.

Smith, S. W. (1997). *The scientist and engineer's guide to digital signal processing*. California Technical Pub.

Unpingco, J. (2014). *Python for Signal Processing*. Springer.

# Contents

7.4    Exercises .......................................... 136
References .............................................. 137

**8    Parameter Fitting** ..................................... 139
8.1    Correlations........................................ 139
    8.1.1    Correlation Coefficient ...................... 139
    8.1.2    Coefficient of Determination ................. 140
8.2    Straight Lines ..................................... 142
    8.2.1    Normal Form of Line Equation ............... 143
8.3    Line Fitting....................................... 143
    8.3.1    Residuals ................................. 144
    8.3.2    Least Squares Estimators .................... 144
8.4    Linear Fits with Python ............................. 144
    8.4.1    Linear Model without Intercept ............... 144
    8.4.2    Linear Model with Intercept ................. 145
    8.4.3    Line-Fit .................................. 146
    8.4.4    Polynomial-Fit ............................ 146
    8.4.5    Sine-Fit................................... 147
    8.4.6    Circle-Fit ................................ 148
8.5    Confidence Intervals ................................ 149
    8.5.1    Finding Confidence Intervals ................. 149
    8.5.2    Confidence Intervals and Hypothesis Tests ......... 151
    8.5.3    Significance ............................... 151
8.6    Fitting Nonlinear Functions .......................... 152
8.7    Exercises .......................................... 154

**9    Spectral Signal Analysis** ............................... 157
9.1    Transforming Data................................. 157
9.2    Fourier Integral ................................... 159
    9.2.1    Definition and Interpretation ................. 159
    9.2.2    Complex Exponential Notation ................ 160
    9.2.3    Examples ................................. 161
9.3    Fourier Series ..................................... 162
    9.3.1    Definition ................................. 162
    9.3.2    Applications................................ 163
9.4    Discrete/Fast Fourier Transform....................... 164
9.5    Spectral Density Estimation.......................... 169
    9.5.1    Periodogram .............................. 169
    9.5.2    Welch Periodogram ......................... 170
9.6    Fourier Transformation, Convolution,
    and Cross-Correlation .............................. 171
    9.6.1    Convolution ............................... 171
    9.6.2    Cross-Correlation .......................... 172
9.7    Time Dependent Fourier Transform .................... 172
    9.7.1    Windowing................................ 172
    9.7.2    Example: Human Vowels..................... 173
9.8    Exercises .......................................... 175
References .............................................. 176

# Chapter 1

# Introduction

This chapter establishes the basics: it describes the typical workflow in signal processing, the conventions used and the basic mathematical tools needed, and the accompanying software that is provided with this book.

## 1.1 Signal Processing

### 1.1.1 Typical Workflow

The analysis of signals involves much more than the implementation of a computer program. In fact, I recommend to always start out *not* with a computer, but with a pencil and a sheet of paper! This makes it easier to focus on the algorithms and not on the hurdles that invariably come up with the programming implementation. Also, one should keep in mind that the first idea on how to achieve the desired goal is often not the most efficient and/or final solution that will be implemented.

A typical real-life workflow involves the following steps:

- Definition of the problem: One can spend a lot of time on data analysis and/or simulation. But it is much more important to ask the right questions, and to make sure that the problem identified has not already been solved by someone else.

- Identification of the data of interest: For this it is often necessary to run some preliminary experiments, and/or give it a quick first shot at data analysis. Crucial factors often only become obvious once you look at a real set of representative data.

- Data collection

- Data import into the program

- Evaluation of the data quality: Start by visualizing the data. Also, it often really helps to collect data from a paradigm where you know the correct results. For example, if your experiment is about analyzing movements, first move your sensor by a well defined amount forward, left, and up, and check if your analysis program correctly reproduces the magnitude and direction of these movements.

- Preprocessing: In general, manipulate your data as little as possible. But if you are aware of certain problems with the data (e.g. 50/60 Hz noise from the electrical environment), you should eliminate those before the main part of data analysis.

- Creation of artificial data, where you know the answer absolutely, to test your analysis programs.

T. Haslwanter, *An Introduction to Hands-on Signal Analysis with Python*,
https://doi.org/10.1007/978-3-030-57903-6_1

- Application of data analysis to measurements.

- Presentation of results

## 1.2 Conventions and Mathematical Basics

The following conventions will be used:

### 1.2.1 Notation

- Axes indexing starts at *0*, and *(0, 1, 2)* corresponds to the $(x, y, z)$ axes, respectively.

- Scalars are indicated by plain letters (e.g. $a$).

- Column vectors are written with bold lowercase letters (e.g. $\mathbf{r}$ ) or in round brackets, and the components of 3-D coordinate systems are labeled $(x, y, z)$:

$$\mathbf{r} = \begin{pmatrix} r_x \\ r_y \\ r_z \end{pmatrix}$$

- The length or "norm" of a vector is indicated by the same name but in plain style.

$$|\mathbf{r}| = \sqrt{\sum_i r_i^2} = r$$

- Matrices are written with bold uppercase letters (e.g. $\boldsymbol{R}$) or in square brackets.

$$\boldsymbol{R} = \begin{bmatrix} R_{xx} & R_{xy} & R_{xz} \\ R_{yx} & R_{yy} & R_{yz} \\ R_{zx} & R_{zy} & R_{zz} \end{bmatrix}$$

- Vector- and matrix-elements are written in plain style, with indices denoted by subscripts (e.g. $r_x$; $R_{yz}$).

- Multiplications with a scalar are denoted by $*$ (e.g. $\tan(\theta/2) * \mathbf{n}$).

- Scalar vector-products and matrix-multiplications are denoted by $\cdot$ (e.g. $\mathbf{p} \cdot \mathbf{q}$).

- Text that is to be typed in at the computer is written in Courier font, e.g. `plot(x,y)`.

- Optional text in command-line entries is expressed by $< ... >$, e.g. `<InstallationDir>`.

- Names referring to computer programs and applications are written in italics, e.g. *Jupyter*.

- Italics will also be used when introducing new terms or expressions for the first time.

- Really important points that should definitely be remembered are indicated as follows:

This will be important!

### 1.2.2 Mathematical Basics

#### Scalar Product

The scalar product of two vectors **a** and **b** is defined as

$$\begin{pmatrix} a_x \\ a_y \\ a_y \end{pmatrix} \cdot \begin{pmatrix} b_x \\ b_y \\ b_z \end{pmatrix} = a_x b_x + a_y b_y + a_z b_z = |\mathbf{a}| * |\mathbf{b}| * \cos(\theta) \tag{1.1}$$

The scalar product $\mathbf{a} \cdot \mathbf{b}$ corresponds to the multiplication of the length **a** with the projection of **b** onto **a** (Fig. 1.1, left). As a consequence, the scalar product is 0 if the two vectors are perpendicular ($\theta = \pi/2$).

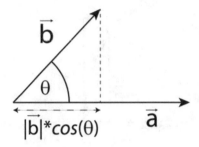

Figure 1.1: Graphical interpretation of scalar-product.

#### Matrix Multiplication

In general, the multiplication of two matrices $\boldsymbol{A}$ and $\boldsymbol{B}$ is defined as

$$\boldsymbol{A} \cdot \boldsymbol{B} = \boldsymbol{C} \tag{1.2}$$

with $C_{ik} = \sum_j A_{ij} B_{jk}$.

To make it easy to remember, think "row * column" (e.g. $C_{12}$ is "row *1* times column *2*"), or remember that one has to sum over adjoining indices.

This equation, which can be represented in Python $\geq$3.5 with the operator "@", can also be used for multiplication of a matrix with a vector when the vector is viewed as a matrix with one column.[1]

Matrix multiplications are implemented in *numpy*, the Python package implementing vector and array calculations (see Fig. 2.2). This can save not only a lot of coding but also a lot of time. For example, in IPython (see Sect. 2.3) the internal matrix multiplication can be compared to a hand-written loop implementation with

```
import numpy as np

A = np.random.randn(500,500)
B = np.random.randn(500,500)

# The IPython magic command "%timeit" measures execution times of commands
%timeit A@B

def calc_me(A, B):
```

---

[1]When working with three- or more-dimensional matrices one has to be careful: the @ operator calls the array's \_\_matmul\_\_ method, not np.dot, and the two methods behave different for > 2 dimensions!

```
    C = np.zeros(A.shape)
    for ii in range(500):
        for jj in range(500):
            for kk in range(500):
                C[ii,jj] += A[ii,kk]*B[kk,jj]
    return C

%timeit calc_me(A,B)
```

On my computer the *numpy* matrix multiplication with "@" is 28'000 times faster than the loop!

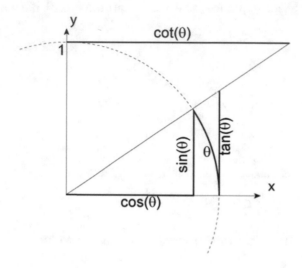

Figure 1.2: Basic trigonometry, with the dashed line representing the unit circle.

**Basic Trigonometry**

The basic elements of trigonometry are illustrated in Fig. 1.2.

**Complex Numbers and Euler's Formula**

Complex numbers have a real part, $a$, and an imaginary part $b$.

With $j = \sqrt{-1}$ the imaginary unit, complex numbers are written as

$$c = a + j * b \tag{1.3}$$

The *complex conjugate* $c^*$ is given by

$$c^* = a - j * b \tag{1.4}$$

Complex numbers are often visualized as points in an x/y-plane, with the real part the x-component, and the imaginary part the y-component. Using this representation, oscillations can be described elegantly with Euler's formula (Fig. 1.3):

$$e^{j\omega t} = \cos(\omega t) + j\sin(\omega t) \tag{1.5}$$

### 1.2.3   Discrete Signals

*Discrete signals* are signals that are sampled at fixed points in time. *Digital signals* are signals that are stored with finite precision. (This can lead to artifacts in data analysis.) In practice, all signals we work with in signal analysis are discrete, digital signals.

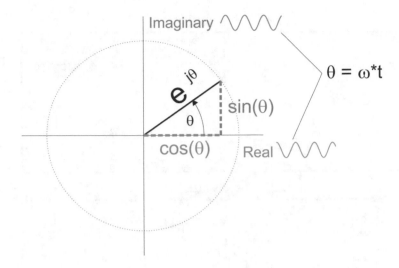

Figure 1.3: The exponential notation is the most elegant way to work with oscillations.

The information obtainable from discrete signals is limited by the *Nyquist theorem*: The highest frequency which can be resolved is half the sampling frequency. Higher frequencies can show up as artifacts (see Fig. 1.4).

## 1.3   Accompanying Material

All the examples and solutions shown in this book are available online. This includes code samples and example programs, Jupyter notebooks with additional or extended information, as well as the data and Python code used to generate many of the figures. They can be downloaded from *github* https://github.com/thomas-haslwanter/sapy, and are organized as follows:

**data** Raw data required for running the programs

**pictures** Images used in this repository

**ipynbs** Jupyter notebooks with additional or extended information relevant to signal processing with Python.

**src/exercise_solutions** Solutions to the exercises that are presented at the end of most chapters

**src/listings** Programs that are explicitly listed in this book

**src/figures** Code used to generate the Python figures in the. Unless noted otherwise, the source code for Python figures is available in the source-file `F[chapter-#]_[figure-#]_xxx.py`. For example, Fig. 1.4 can be generated with the Python code in `F1_4_nyquist.py`

**src/code-quantlets** Additional code samples. Figures that have been generated by such code-quantlets refer to them with "CQ [name of quantlet]". For example, Fig. 5.24 refers to "CQ `bspline_demo.py`"

**tests** Test functions for the code in **src**

Make sure to look at the file `Errata.pdf`, which will be kept up-to-date with corrections to any mistakes that are discovered after publication of the book. If you find any new errors, please report them at https://github.com/thomas-haslwanter/sapy/issues.

Code samples that are independent from figures in the text are marked as follows:

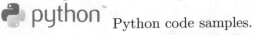 Python code samples.

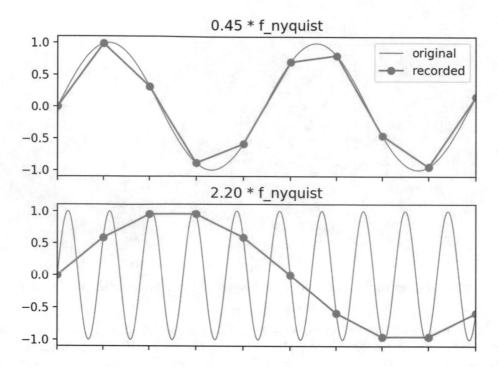

Figure 1.4: **Top:** When the signal frequency is below the Nyquist frequency, the important signal features are preserved. **Bottom:** When it is above the Nyquist frequency, the recorded data are aliased, important signal components are lost, and artifacts are introduced.

Packages on *github* are called *repositories*, and can easily be copied to your computer: when git is installed on your computer, simply type `git clone` https://github.com/thomas-haslwanter/sapy.git (here the repository name of `signal-processing` given above) in a command terminal and the whole repository—code as well as data—will be "cloned" to your system. (See Sect. 12.2 for more information on *git*, *github* and code-versioning.) Alternatively, you can download a ZIP-archive from there to your local system.

## 1.4   Exercises

1. **Matrix Multiplication**
   Take a sheet of paper and try to write down the solutions to

   (a) $\begin{pmatrix} 1 & 2 \end{pmatrix} \cdot \begin{pmatrix} 3 \\ 4 \end{pmatrix}$

   (b) $\begin{pmatrix} 3 \\ 4 \end{pmatrix} \cdot \begin{pmatrix} 1 & 2 \end{pmatrix}$

   (c) $\begin{pmatrix} 1 & 2 \end{pmatrix} \cdot \begin{pmatrix} 3 & 4 \end{pmatrix}$

   When you are done with that, try to execute them in Python and check your results.

# Chapter 2

# Python

Python is free, consistently and completely object oriented, and has a large number of (free) scientific toolboxes (e.g. http://www.scipy.org/). It is used by Google, NASA, and many others. Information on Python can be found under http://www.python.org/. If you want to use Python for scientific applications, currently the best way to get started is with a Python distribution, either *WinPython*, or Anaconda from *Continuum Analytics*. These distributions are free and contain the complete scientific and engineering development software for numerical computations, data analysis and data visualization based on Python. They also come with *Qt* graphical user interfaces, and the interactive scientific/development environment Spyder. If you already have experience with *Matlab*, the article *NumPy for Matlab Users* (https://numpy.org/devdocs/user/numpy-for-matlab-users.html) provides an overview of the similarities and differences between the two languages.

Figure 2.1: The Python Logo.

Python is a very-high-level dynamic object-oriented programming language (Fig. 2.1). It is designed to be easy to program and easy to read. It was started in 1980, and has since gained immense popularity in a broad range of fields from web development, system administration, and in science and engineering. Python is open source, and has become one of the most successful programming languages. There are three reasons why I have switched from other programming languages to Python:

1. It is the most elegant programming language that I know.

2. It is free.

3. It is powerful.

## 2.1 Getting Started

### 2.1.1 Distributions and Packages

The Python core distribution contains only the essential features of a general programming language. Figure 2.2 shows the modular structure of the most important Python packages that are used in this book. Python itself is an interpretative programming language, with no optimization for working with vectors or matrices, or for producing plots. *Packages* which extend the abilities of Python must be loaded explicitly. The most important packages for scientific

T. Haslwanter, *An Introduction to Hands-on Signal Analysis with Python*,
https://doi.org/10.1007/978-3-030-57903-6_2

applications are *numpy*, which makes working with vectors and matrices fast and efficient, and *matplotlib*, which is the most common package used for producing graphical output. *scipy* contains important scientific algorithms. And *pandas* has become widely adopted for statistical data analysis. It provides *DataFrames* which are labeled, 2-dimensional data structures, making work with data more flexible and intuitive.

*IPython* provides the tools for interactive data analysis, and *Jupyter* provides the different front-ends for *IPython*. *IPython* lets you quickly display graphs and change directories, explore the workspace, provides a command history etc.

To facilitate the use of Python, so called *Python distributions* collect matching versions of the most important packages, and I strongly(!) recommend using one of these distributions when getting started. Otherwise one can easily become overwhelmed by the huge number of Python packages available. My favorite Python distributions are

- *WinPython* recommended for Windows users. At the time of writing, the latest version was 3.9.2
  https://winpython.github.io/

- *Anaconda* by Continuum. For Windows, Mac, and Linux. The latest *Anaconda* version at the time of writing was 2020.11, with Python 3.8.
  https://www.anaconda.com/products/individual

I am presently using *WinPython*, which is free and customizable. *Anaconda* also runs under Mac OS and under Linux, and is free for educational purposes.

The Python code samples in this book expect a Python version $\geq$3.6.

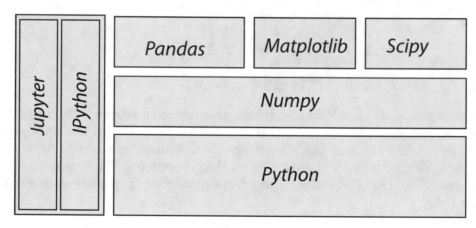

Figure 2.2: The structure of the most important Python packages for signal processing.

The programs included in this book have been tested under Windows (Python 3.8.8) and Linux (Python 3.7.4). The numbers of the programs tested are:

- *Python 3.8.8* ... Basic Python installation,

- *ipython 7.21.0* ... For interactive work.

- *numpy 1.20.1+mkl* ... For working with vectors and arrays.

- *scipy 1.6.1* ... All the essential scientific algorithms

- *matplotlib 3.3.4* ... The de-facto standard module for plotting and visualization.

- *pandas 1.2.3* ... Adds *DataFrames* (imagine powerful spreadsheets) to Python.

- *Jupyter 1.0.0* ... For interactive work environments, e.g. the *JupyterLab*, *Jupyter Notebook*, or the *Qt Console*

All of these packages come with the *WinPython* and *Anaconda* distributions. Additional packages, which may be required by individual applications, can easily be installed using `pip` or `conda`.

**PyPI—The Python Package Index**

The Python Package Index (*PyPI*) (https://pypi.org/) is a repository of software for the Python programming language and contains more than 300'000 projects!

Packages from *PyPI* can be installed easily from the Windows command shell (`cmd`) or the Linux `terminal` with

```
pip install <package>
```

To update a package, use

```
pip install <package> -U
```

To get a list of all the Python packages installed on your computer, type

```
pip list
```

*Anaconda* uses `conda`, a more powerful installation manager. But *pip* also works with *Anaconda*.

## 2.1.2 Installation of Python

While Python and the required packages can be installed manually, it is typically much easier to start out with a complete Python distribution.

**Under Windows**

Neither *WinPython* nor *Anaconda* require administrator rights for installation.

**WinPython**  In the following, I assume that `<WinPythonDir>` is the installation directory for *WinPython*.

**Tip:** Do NOT install *WinPython* into the Windows program directory (typically `C:\Program Files` or `C:\Program Files (x86)`), because this can lead to permission problems during the execution of *WinPython*.

- Download WinPython from https://winpython.github.io/.

- Run the downloaded `.exe`-file, and install *WinPython* into the `<WinPythonDir>` of your choice. (On my own system I place all programs that do not modify the Windows Registry, such as *WinPython*, *vim*, *ffmpeg* etc. into a folder `C:\Programs`.)

- After the installation, make a change to your *Windows Environment*, by typing `Win -> env -> Edit environment variables for` *your* `account` (Note that this is different from the `system environment`!):

- – Add the directories
  `<WinPythonDir>\python-3.8.1.amd64;`
  `<WinPythonDir>\python-3.8.1.amd64\Scripts\;`
  to your `PATH`. (This makes Python and *IPython* accessible from the standard *Windows* command-line, which can be reached quickly by typing `Win+cmd`.)
- – Remove the default `%USERPROFILE%\AppData\Local\Microsoft\Windows` `Apps` from the `PATH` (since it contains a misleading `python.exe`-link).
- – If you do have administrator rights, you should activate
  `<WinPythonDir>\WinPython Control Panel.exe ->`
  `Advanced -> Register Distribution.`
  (This associates `.py`-files with this Python distribution.)

**Anaconda**

- Download *Anaconda* from https://www.anaconda.com/distribution/.

- Follow the installation instructions from the web page. During the installation, allow *Anaconda* to make the suggested modifications to your environment `PATH`.

- After the installation: in the *Anaconda Launcher*, click *update* (besides the apps), in order to ensure that you are running the latest version.

**Installing additional packages** When I have had difficulties installing additional packages, I have been saved more than once by the pre-compiled packages Christoph Gohlke, available under http://www.lfd.uci.edu/~gohlke/pythonlibs/: from there you can download the `[xxx].whl` file for your current version of Python, and then install it simply with `pip install [xxx].whl`.

**Under Linux**

The following procedure works on *Linux Mint 20.1*:

- Download the most recent version of *Anaconda*.

- Open `terminal`, and navigate to the location where you downloaded the file to.

- Install *Anaconda* with `bash Anaconda[xx]-y.y.y-Linux-x86.sh`

- Update your Linux installation with `sudo apt-get update`

**Notes**

- You do not need root privileges to install *Anaconda* if you select a user writable install location, such as `~/Anaconda`.

- After the self extraction is finished, you should add the *Anaconda* binary directory to your `PATH` environment variable.

- As all of *Anaconda* is contained in a single directory, uninstalling *Anaconda* is easy: you simply remove the entire install location directory.

- If any problems remain, Mac and Unix users should look up Johansson' installations tips: https://github.com/jrjohansson/scientific-python-lectures.

**Under Mac OS X**

- Go to https://www.anaconda.com/distribution/

- Choose the Mac installer (make sure you select the *Mac OS X Python 3.x Graphical Installer*, and follow the instructions listed beside this button.

- After the installation: in the *Anaconda Launcher*, click *update* (besides the Apps), in order to ensure that you are running the latest version.

After the installation the *Anaconda* icon should appear on the desktop. No admin password is required. This downloaded version of *Anaconda* includes the *Jupyter notebook*, *Jupyter QtConsole* and the IDE *Spyder*.

To see which packages (e.g. *numpy, scipy, matplotlib, pandas*, etc.) are featured in your installation look up the *Anaconda Package List* for your Python version. For example, the Python-installer may not include *seaborn*. To add an additional package, e.g. *seaborn*, open the terminal, and enter `pip install seaborn`.

### 2.1.3   Python Resources

My favorite introductory book for scientific applications of Python is Scopatz and Huff [2015]. However, that book does not provide any information on signal processing. If you have some programming experience, the book you are currently reading may be all you need to get the signal analysis of your data going. If required, very good additional information can be found on the web, where tutorials as well as good free books are available online. The following links are all recommendable sources of information for starting with Python:

- *Python Scientific Lecture Notes* If you don't read anything else, read this! (http://scipy-lectures.org/)

- *NumPy for Matlab Users* Start here if you have Matlab experience. (https://numpy.org/doc/stable/user/numpy-for-matlab-users.html; also check http://mathesaurus.sourceforge.net/matlab-numpy.html)

- *Lectures on scientific computing with Python* Great *Jupyter notebooks*, from JR Johansson. (https://github.com/jrjohansson/scientific-python-lectures)

- *The Python tutorial* The official introduction. (http://docs.python.org/3/tutorial)

When running into a problem while developing a new piece of code, most of the time I just google; thereby I stick primarily to the official Python documentation pages and to http://stackoverflow.com. Also, I have found Python user groups surprisingly active and helpful!

### 2.1.4   A Simple Python Program

**Hello World**

**Python Shell** The simplest way to start Python is to type `python` on the command line. (When I say *command line* I refer in *Windows* to the command shell started with `cmd`, and in *Linux* or *Mac OS X* to the `terminal`.) Then you can already start to execute Python commands, e.g. the command to print "Hello World" to the screen: `print('Hello World')`. On my Windows computer, this results in

```
Python 3.8.8 (tags/v3.8.8:024d805, Feb 19 2021, 13:18:16)
[MSC v.1928 64 bit (AMD64)] on win32
Type "help", "copyright","credits" or "license" for more information.
>>> print('Hello World')
Hello World
>>>
```

However, most of the time it is more recommendable to start with the *IPython/Jupyter* Qt Console described in more detail in Sect. 2.3. The Qt-console is an interactive programming environment which offers a number of advantages. For example, when you type `print(` in the Qt console, you immediately see information about the possible input arguments for the command `print`.

**Python Modules** are files with the extension `.py`, and are used to store Python commands in a file for later use. Let us create a new file with the name `helloWorld.py`, which contains the line

```
print('Hello World')
```

This file can now be executed by typing `python helloWorld.py` on the command line.

On *Windows* you can actually run the file by double-clicking it, or by simply typing `helloWorld.py`, if the extension `.py` is associated with the local Python installation. On *Linux* and *Mac OS X* the procedure is slightly more involved. There, the file needs to contain an additional first line specifying the path to the Python installation.

```
#! \usr\bin\python
print('Hello World')
```

On these two systems you also have to make the file executable, by typing

```
chmod +x helloWorld.py
```

before you can run it with `helloWorld.py`.


**square_me**

To increase the level of complexity, let us write a Python module that includes a function definition and prints out the square of the numbers from zero to five. We call the file `L2_1_square_me.py`, and it contains the following lines

```
Listing 2.1: square_me.py
```

```
1  # This file shows the square of the numbers from 0 to 5.
2
3  def squared(x=10):
4      return x**2
5
6  for ii in range(6):
7      print(ii, squared(ii))
8
9  print( squared() )
```

Let me explain what happens in this file, line-by-line:

**1** The first line starts with "#", indicating a comment-line.

**3–4** These two lines define the function `squared`, which takes the variable $x$ as input, and returns the square (`x**2`) of this variable. If the function is called with no input, $x$ is by default set to *10*. This notation makes it very simple to define default values for function inputs.

**Note:** The range of the function is defined by the indentation! This is a feature loved by many Python programmers, but often found a bit confusing by newcomers. Here the last indented line is *line 4*, which ends the function definition.

**6–7** Here the program loops over the first 6 numbers. Also the range of the `for`-loop is defined by the indentation of the code.

In *line 7*, each number and its corresponding square are printed to the output.

**9** This command is not indented, and therefore is executed after the `for`-loop has ended. It tests if the function call with "()", which uses the default parameter for $x$, also works, and prints the result.

**Notes:**

- Since Python starts at 0, the loop in *line 6* includes the six numbers from 0 to 5.

- In contrast to some other languages Python distinguishes the syntax for function calls from the syntax for addressing elements of an array etc: function calls, as in *line 7*, are indicated with round brackets ( ... ); and individual elements of arrays or vectors are addressed by square brackets [ ... ].

## 2.2 Elements of Scientific Python Programming

### 2.2.1 Python Datatypes

Python offers a number of powerful data structures, and it pays off to make yourself familiar with them. The most common ones are:

- *Lists* to group objects of the same types.

- *Numpy Arrays* to work with numerical data. (numpy also offers the data type `np.matrix`. However, in my experience `np.array` is the way to go, since many numerical and scientific functions will not accept input data in `matrix` format.)

- *Tuples* to group objects of different types.

- *Dictionaries* for named, structured data sets.

- *Pandas DataFrames* for simple import and export of data, and for statistical data analysis.

For simple programs you will mainly work with lists and arrays. Dictionaries are used to group related information together. And tuples are used primarily to return multiple parameters from functions.

**List [ ]** Lists are typically used to collect items of the same type (numbers, strings, ...). They are "mutable", i.e. their elements can be modified.

Note that "+" concatenates lists.

```
In [4]: myList = ['abc', 'def', 'ghij']

In [5]: myList.append('klm')
```

```
In [6]: myList
Out[6]: ['abc', 'def', 'ghij', 'klm']

In [7]: myList2 = [1,2,3]

In [8]: myList3 = [4,5,6]

In [9]: myList2 + myList3
Out[9]: [1, 2, 3, 4, 5, 6]
```

**Array [ ]** *vectors* and *matrices*, for numerical data manipulation. Defined in *numpy*. Note that vectors and 1-d arrays are different: vectors CANNOT be transposed! With arrays, "+" adds the corresponding elements; and the array-method *.dot* performs a scalar multiplication. (Since Python 3.5, scalar multiplications can also be performed with the operator "@".)

```
In [10]: myArray2 = np.array(myList2)

In [11]: myArray3 = np.array(myList3)

In [12]: myArray2 + myArray3
Out[12]: array([5, 7, 9])

In [13]: myArray2.dot(myArray3)
Out[13]: 32

In [14]: myArray2 @ myArray3
Out[14]: 32
```

**Tuple ( )** A collection of different things. Once created tuples cannot be modified. (This really irritated me when I started to work with Python. But since I use tuples almost exclusively to return parameters from functions, this has not turned out to be any real limitation.)

```
In [1]: import numpy as np

In [2]: myTuple = ('abc', np.arange(0,3,0.2), 2.5)

In [3]: myTuple[2]
Out[3]: 2.5
```

**Dictionary { }** Dictionaries are unordered *(key/value)* collections of content, where the content is addressed as dict['key']. Dictionaries can be created with the command dict, or by using curly brackets {...}:

```
In [15]: myDict = dict(one=1, two=2, info='some information')

In [16]: myDict2 = {'ten':1, 'twenty':20,
'info':'more information'}

In [17]: myDict['info']
Out[17]: 'some information'

In [18]: myDict.keys()
Out[18]: dict_keys(['one', 'info', 'two'])
```

**DataFrame** Data structure optimized for working with named, statistical data. Defined in *pandas*. (See Sect. 2.2.4.)

### 2.2.2 Indexing and Slicing

The rules for addressing individual elements in Python lists, tuples, or *numpy* arrays have been nicely summarized by Greg Hewgill on *stackoverflow*[1]:

```
a[start:end] # items start through end-1
a[start:]    # items start through the rest of the array
a[:end]      # items from the beginning through end-1
a[:]         # a copy of the whole array
```

There is also the `step` value, which can be used, for example the above:

```
a[start:end:step] # start through not past end, by step
```

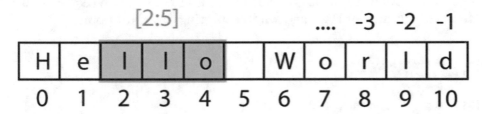

Figure 2.3: Indexing starts at 0, and slicing does *not* include the last value.

The key points to remember are that indexing starts at 0, *not* at 1; and the `:end` value represents the first value that is *not* in the selected slice. So, the difference `end - start` is the number of elements selected (if `step` is 1, the default).

`start` or `end` may be a negative number. In that case the count goes from the end of the array instead of the beginning. So:

```
a[-1]    # last item in the array
a[-2:]   # last two items in the array
a[:-2]   # everything except the last two items
```

As a result, `a[:5]` gives you the first five elements (*Hello* in Fig. 2.3), and `a[-5:]` the last five elements (*World*).

### 2.2.3 Numpy Vectors and Arrays

*numpy* is the Python module that makes working with numbers efficient. It is commonly imported with

```
import numpy as np
```

By default, it produces vectors. The commands most frequently used to generate numbers are:

**np.zeros** generates numpy arrays containing zeros. Note that it takes only one(!) input. If you want to generate a matrix of zeroes, this input has to be a tuple or a list, containing the number of rows/columns!

---

[1]http://stackoverflow.com/questions/509211/explain-pythons-slice-notation.

```
In [1]: import numpy as np

In [2]: np.zeros(3) # by default numpy-functions generate 1D-vectors
Out[2]: array([ 0.,   0.,   0.])

In [3]: np.zeros( (2,3) )
Out[3]: array([[ 0.,   0.,   0.],
               [ 0.,   0.,   0.]])
```

**np.ones** generates numpy arrays containing ones.

**np.random.randn** generates normally distributed numbers, with a mean of *0* and a standard deviation of *1*. To produce reproducible random numbers you have to specify the starting point for the random number generation, for example with `np.random.seed(...)`, using an integer number of your choice.

**np.arange** generates a range of numbers. Parameters can be `start, end, steppingInterval`. Note that the end-value is excluded! While this can sometimes be a bit awkward, it has the advantage that consecutive sequences can be easily generated, without any overlap, and without missing any data points:

```
In [4]: np.arange(3)
Out[4]: array([0, 1, 2])

In [5]: xLow = np.arange(0, 3, 0.5)
In [6]: xHigh = np.arange(3, 5, 0.5)

In [7]: xLow
Out[7]: array([ 0., 0.5, 1., 1.5, 2., 2.5])

In [8]: xHigh
Out[8]: array([ 3., 3.5, 4., 4.5])
```

**np.linspace** generates linearly spaced numbers

```
In [9]: np.linspace(0, 10, 6)
Out[9]: array([ 0., 2., 4., 6., 8.,  10.])
```

**np.array** generates a numpy array from given numerical data, and is a convenient notation to enter small matrices

```
In [10]: np.array([[1,2], [3,4]])
Out[10]: array([ [1, 2],
                 [3, 4] ])
```

There are a few points that are peculiar to Python, and that are worth noting:

**Matrices** are simply "lists of lists". Therefore the first element of a matrix gives you the first row, the second element the second row, etc.:

```
In [11]: Amat = np.array([ [1, 2],
                           [3, 4] ])

In [12]: Amat[0]
Out[12]: array([1, 2])
```

**Warning:** A vector is not the same as a 1-dimensional matrix! This is one of the few features of Python that is not intuitive (at least to me), and can lead to mistakes that are hard to find. For example, vectors cannot be transposed, but matrices can.

```
In [13]: x = np.arange(3)

In [14]: Amat = np.array([ [1,2], [3,4] ])

In [15]: x.T == x
Out[15]: array([ True,  True,  True])
# This indicates that a vector stays a vector, and that
# the transposition with ".T" has no effect on its shape

In [16]: Amat.T == Amat
Out[16]: array([[ True, False],
                [False,  True]])
```

**np.r_** Useful command to quickly build up small row-vectors. But I only use it to try things out quickly. I my programs I prefer the clearer but equivalent `np.array([...])`

```
In [17]: np.r_[1,2,3]
Out[17]: array([1, 2, 3], dtype=int32)
```

**np.c_** Useful command to quickly build up small column-vectors. Note that column-vectors can also be generated with the command `np.newaxis`:

```
In [18]: np.c_[[1.5,2,3]]   # note the double square brackets!
Out[18]:
array([[1.5],
       [2. ],
       [3. ]])

In  [19]: x[:, np.newaxis]
Out [19]:
array([[0],
       [1],
       [2]])
```

**np.atleast_2d** converts a vector (which cannot be transposed, see above) to the corresponding 2-dimensional array (which can be transposed):

```
In [20]: x = np.arange(5)

In [21]: x
Out[21]: array([0, 1, 2, 3, 4])

In [22]: x.T
Out[22]: array([0, 1, 2, 3, 4])    # no effect on 1D-vectors

In [23]: x_2d = np.atleast_2d(x)

In [24]: x_2d.T
Out[24]:
array([[0],
       [1],
       [2],
       [3],
       [4]])
```

**np.column_stack** An elegant command to generate column-matrices:

```
In [25]: x = np.arange(3)

In [26]: y = np.arange(3,6)

In [27]: np.column_stack( (x,y) )
Out[27]:
array([[0, 3],
       [1, 4],
       [2, 5]])
```

### 2.2.4   Pandas DataFrames

*pandas* (http://pandas.pydata.org/) is a widely used Python package, and provides data structures suitable for statistical analysis and data manipulation. It also adds functions that facilitate data input, data organization, and data manipulation. *pandas* is commonly imported with

```
import pandas as pd.
```

The official pandas documentation contains a very good "Getting started" section: https://pandas.pydata.org/docs/getting_started/.

#### Basic Syntax of DataFrames

Especially in statistical data analysis (read "data science"), *Pan*elled *da*ta structures (→ "Pandas") have turned out to be immensely useful. To handle such labeled data in Python, *pandas* introduces so-called *DataFrame* objects. A DataFrame is a 2-dimensional labeled data structure with columns of potentially different types. You can think of it like a spreadsheet or SQL table (see Fig. 2.4). DataFrames are the most commonly used *pandas* objects.

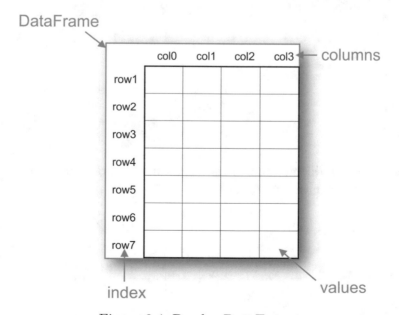

Figure 2.4: Pandas DataFrame.

For statistical analysis, *pandas* becomes really powerful when combined with the package *statsmodels* (https://www.statsmodels.org/).

*Pandas* DataFrames can have some distinct advantages over *numpy* arrays:

- A numpy array requires homogeneous data. In contrast, with a pandas DataFrame you can have a different data type (float, int, string, datetime, etc) in each column (Fig. 2.5).

- Pandas has built in functionality for a lot of common data-processing applications: for example, easy grouping by syntax, easy joins (which are also really efficient in pandas), rolling windows, etc.

- DataFrames, where the data can be addressed with column names, can help a lot in keeping track of your data.

In addition, *pandas* has excellent tools for data input and output.

Let me start with a specific example, by creating a DataFrame with three columns, called "Time", "x", and "y":

```
import numpy as np
import pandas as pd

t = np.arange(0, 10, 0.1)
x = np.sin(t)
y = np.cos(t)

df = pd.DataFrame({'Time':t, 'x':x, 'y':y})
```

In Pandas, rows are addressed through indices, and columns through their name. To address the first column only, you have two options:

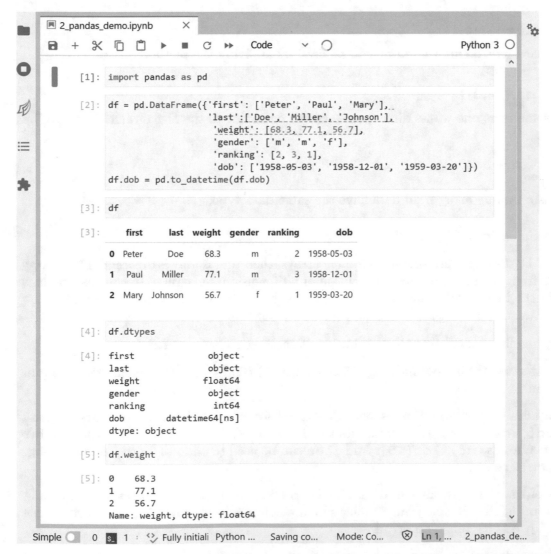

Figure 2.5: Demonstration of some features of pandas DataFrames in *JupyterLab* (see Sect. 2.3.1). Unlike `np.array`, a `pd.DataFrame` can be used to combine different data types.

```
df.Time
df['Time']
```

To extract two columns at the same time, put the variable names in a list. With the following command, a new DataFrame `data` is generated, containing the columns `Time` and `y`:

```
data = df[['Time', 'y']]
```

After reading in the data it is good practice to check if the data have been read in correctly. The first or last few rows can be displayed with:

```
data.head()
data.tail()
```

For example, the following statement shows rows 5–10 (note that these are 6 rows):

```
data[4:10]
```

as $10 - 4 = 6$. (I know, the array indexing takes some time to get used to. It helps me to think of the indices as pointers *to* the elements, and that they start at 0.)

The handling of DataFrames is somewhat different from the handling of *numpy* arrays. For example, (numbered) rows and (labeled) columns can be addressed simultaneously as follows:

```
df[['Time', 'y']][4:10]
```

The standard row/column notation can be used by applying the method `iloc`:

```
df.iloc[4:10, [0,2]]
```

Finally, sometimes one wants direct access to the data, not to the DataFrame. This can be achieved with

```
data.values
```

which returns a *numpy* array if all data have the same data type.

**Note: Data Selection**

While *pandas'* DataFrames are similar to *numpy* arrays, their philosophy is different. The *numpy* syntax comes from the mathematical description of n-dimensional matrices. In contrast, *pandas* has its origin the in the data analysis of column oriented data base information. Some of the differences between the two that you should watch out for are:

- *numpy* handles "rows" first. E.g. `data[0]` is the first row of an array

- *pandas* starts with the columns. E.g. `df['values'][0]` is the first element of the column `'values'`.

- If a DataFrame has labeled rows, one can extract for example the row "row_label" with `df.loc['row_label']`. If one wants to address a row by its number, e.g. row number "15", one can use `df.iloc[15]`. And `iloc` can be used to address "rows/columns", e.g. `df.iloc[2:4,3]`.

- Slicing of rows also works, e.g. `df[0:5]` for the first 5 (!) rows. A sometimes confusing convention is that if you want to slice out a single row, e.g. row "5", you have to use `df[5:6]`. `df[5]` raises an error!

 python **Code:** `show_pandas.py` (Appendix A) demonstrates some of *pandas* powerful functions to handle missing data, and for "grouping" of data items.

### 2.2.5 Python Documentation

You should ALWAYS document the code that you write—even if you only hack a small program! I have been surprised how often I have had to go back and modify code that I thought I would "never need again". And how often I then had a hard time to understand my own code if there were no comments included. The complete overview over the recommended best-practices in Python can be found under https://pep8.org/. A short application example is given here in Listing 2.2. Note that in the function definition so-called "type hints" are used, to indicate input and return type. Those are optional, but make the code easier to read and understand.

Listing 2.2: python_module.py

```python
1 """ Demonstration of a Python Function """
2
3 # author:    Thomas Haslwanter
4 # date:      June-2020
5
6 # Import standard packages
7 import numpy as np
8 from typing import Tuple
9
10
11 def income_and_expenses(data : np.ndarray) -> Tuple[float, float]:
12     """Find the sum of the positive numbers, and the sum of the negative ones.
13
14     Parameters
15     ----------
16     data : numpy array (,n)
17             Incoming and outgoing account transactions
18
19     Returns
20     -------
21     income : Sum of the incoming transactions
22     expenses : Sum of the outgoing account transactions
23     """
24
25     income = np.sum(data[data>0])
26     expenses = np.sum(data[data<0])
27
28     return (income, expenses)
29
30
31 if __name__=='__main__':
32     testData = np.array([-5, 12, 3, -6, -4, 8])
33
34     # If only real banks would be so nice ;)
35     if testData[0] < 0:
36         print('Your first transaction was a loss, and will be dropped.')
37         testData = np.delete(testData, 0)
38     else:
39         print('Congratulations: Your first transaction was a gain!')
40
41     (my_income, my_expenses) = income_and_expenses(testData)
42     print(f'You have earned {my_income:5.2f} EUR, and spent {-my_expenses:5.2f}
          EUR.')
```

A detailed description of what happens in this piece of code is given below.

### 2.2.6 Functions, Modules, and Packages

Python has three different levels of modularization:

**Function** is defined by the keyword `def`, and can be defined anywhere in Python. It returns the object in the `return` statement, typically at the end of the function.


**Modules** are files with the extension ".py". Modules can contain function and variable definitions, as well as valid Python statements.

**Packages** are folders containing multiple Python modules, and must contain a file named `__init__.py`. For example, *numpy* is a Python package. Since packages are mainly important for grouping a larger number of modules, they won't be discussed in this book.

### Functions

A function is a set of statements that take inputs, do some specific computation and produce output. The idea is to group commonly or repeatedly done task and make a function, so that instead of writing the same code again and again for different inputs we can call the function. In Python, functions can be declared at any point in a program with the command `def`.

The example in Listing 2.2 shows how functions can be defined and used.

- **1:** Module header.

- **3/4:** Author and date information (should be separate from the module header).[2]

- **6–8:** Since *numpy* will be required in that module, it has to be imported. To reduce the writing to a minimum, it is conventionally called `np`. The command `Tuple` from the package *typing* will be used in the "type hints" for the upcoming function. Type hints give hints on the type of the object(s) the function is using and for its return. They are optional, but improve the readability of code.

- **9/10:** Keep 2 empty lines before function definitions

- **11:** Function signature

- **12–23:** Comment describing the function. Should also include information about the parameters the function takes, and about the return elements.

- **25–28:** Function definition. Note that in Python the function block is defined by the indentation, not by any brackets or *end* statements! This is a feature that irritates many Python novices, but really helps to keep code clear and nicely formatted. Important: Python makes a difference between a tab and the equivalent amount of spaces. This can lead to errors which are really hard to detect, so use a good IDE that automatically converts tabs to spaces!

- **25:**

  - The `sum` command is taken from *numpy*, so it has to be preceded by `np`.
  - In Python, function arguments are indicated by round brackets ( ... ), whereas elements of lists, tuples, vectors, and arrays are indicated by square brackets [ ... ].
  - In *numpy* you can select elements of an array either with an index (see line **35**), or with a Boolean array (lines **25–26**).

- **28:** Python also uses round brackets to form groups of elements, so-called *tuples*. And the `return` statement does the obvious things: it returns elements from a function.

---

[2]For the rest of the book the "author/date" information will be left away, to keep the program listings more compact.

- **31:** Here quite a few new aspects of Python are introduced:

  - Just like function definitions, `if`-loops or `for`-loops use indentation to define their context.

  - A convention followed by most Python coders is to prefix variables or methods that are supposed to be treated as a non-public part of the Python code with an underscore, for example `_geek` or `__name__`.

  - Here we check the variable with the name `__name__`, which is automatically generated by the Python interpreter and indicates the context of a module evaluation. If the module is run as a Python script, `__name__` is automatically set to `__main__`. But if a module is imported (see e.g. Listing 2.3), it is set to the name of the importing module. This way it is possible to add code to a function that is only used when the module is executed, but not when the functions in this module are `imported` by other modules (see below). This is a nice way to test functions defined in the same module.

- **32:** Definition of a *numpy* array.

- **41:** The two elements returned as a tuple from the function `income_and_expenses` can be immediately assigned to two different Python variables, here to (`my_income`, `my_expenses`).

- **42:** While there are different ways to produce formatted strings, the "f-strings" that were introduced with Python 3.6 are probably the most elegant: curly brackets {} indicate values that will be inserted. The optional expression after the colon contains formatting statements: here `:5.2f` indicates "express this number as a float, with 5 digits, 2 of which are after the comma".[3] The corresponding values are then passed into the f-string for formatted output.

### Modules

To execute the module `L2_2_python_module.py` from the command-line, type `python L2_2_python_module.py`. In Windows, if the extension ".py" is associated with the Python program, it suffices to double-click the module, or to type `python_module.py` on the command-line. In *WinPython* the association of the extension ".py" with the Python function can be set by the *WinPython Control Panel.exe*, by the command *Register Distribution ...* in the menu *Advanced*.

To run a module in *IPython*, use the magic function `%run`

```
In [56]: %run L2_2_python_module
Your first transaction was a loss, and will be dropped.
You have earned 23.00 EUR, and spent 10.00 EUR.
```

Note that you either have to be in the directory where the function is defined, or you have to give the full path name.

If you want to use a function or variable that is defined in a different module, you have to import that module. This can be done in three different ways. For the following example, assume that the other module is called `new_module.py`, and the function that we want from there `new_function`.

- `import new_module`: The function can then be accessed with `new_module.new_function()`.

---

[3]https://pyformat.info/ contains all the details of formatted output in Python.

- `from new_module import new_function`: In this case, the function can be called directly `new_function()`.

- `from new_module import *`: This imports all variables and functions from `new_module` into the current workspace; again, the function can be called directly with `new_function()`. However, use of this syntax is discouraged! It clutters up the current workspace, and one risks overwriting existing variables with the same name as an imported variable.

If you import a module multiple times, Python recognizes that the module is already known and skips later imports.

The next example shows you how to import functions from one module into another module:

**Listing 2.3: python_import.py**

```
1  """ Demonstration of importing a Python module """
2
3  # Import standard packages
4  import numpy as np
5
6  # additional packages: this imports the function defined above
7  import L2_2_python_module as py_func
8
9  # Generate test-data
10 testData = np.arange(-5, 10)
11
12 # Use a function from the imported module
13 out = py_func.income_and_expenses(testData)
14
15 # Show some results
16 print(f'You have earned {out[0]:5.2f} EUR, and spent {-out[1]:5.2f} EUR.')
```

- **7:** The module `L2_2_python_module` (that we have just discussed above) is imported, as `py_func`.

- **13:** To access the function `income_and_expenses` from the module `py_func`, module- and function-name have to be given: `py_func.income_and_expenses(...)`

## 2.3 IPython/Jupyter—An Interactive Programming Environment

### 2.3.1 Overview

#### IPython

A good workflow for source code development can make a very big difference for coding efficiency. For me, the most efficient way to write new code is as follows: After figuring out all the algorithms on paper, I first get the individual steps worked out interactively in *IPython* (http://ipython.org/). *IPython* provides a programming environment that is optimized for interactive computing with Python, similar to the command-line in Matlab. It comes with a command history, interactive data visualization, command completion, and a lot of features that make it quick and easy to try out code.

While IPython can also be run in a terminal-environment, its full power becomes available with *Jupyter*.

**Jupyter**

In 2013 the *IPython Notebook*, a browser-based front-end for Python, became a very popular way to share research and results in the Python community. In 2015 the development of the front-end became its own project, called *Project Jupyter* (https://jupyter.org/). Today *Jupyter* can be used not only with Python, but also with *Julia, R*, and more than 100 other programming languages.

The most important interfaces provided by Jupyter are

- *Qt Console*
- *Jupyter Notebook*
- *JupyterLab*

They can be started from a terminal with the command

```
jupyter [viewer]
```

where `viewer` is `qtconsole`, `notebook`, or `lab`.

**Qt Console**

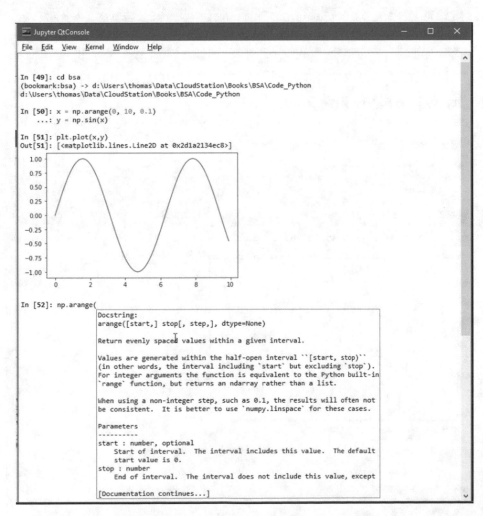

Figure 2.6: The Qt Console, displaying parameter tips for the current command.

The *Qt Console* (see Fig. 2.6) is my preferred way to start coding, especially to figure out the correct command syntax. It provides immediate feedback on the command syntax, and good text completion for commands, file names, and variable names.

**Jupyter Notebook**

The Jupyter Notebook is a browser based interface, which is especially well suited for teaching, documentation, and for collaborations. It allows you to combine a structured layout, equations in the popular LaTeX-format, and images, and can include resulting graphs and videos, as well as the output from Python commands (see Fig. 2.7). Packages such as *plotly* (https://plot.ly/) or *bokeh* (https://bokeh.org/) build on such browser-based advantages, and allow easy construction of interactive interfaces inside Jupyter Notebooks.

Code samples accompanying this book are also available as *Jupyter Notebooks*, and can be downloaded from https://github.com/thomas-haslwanter/sapy.

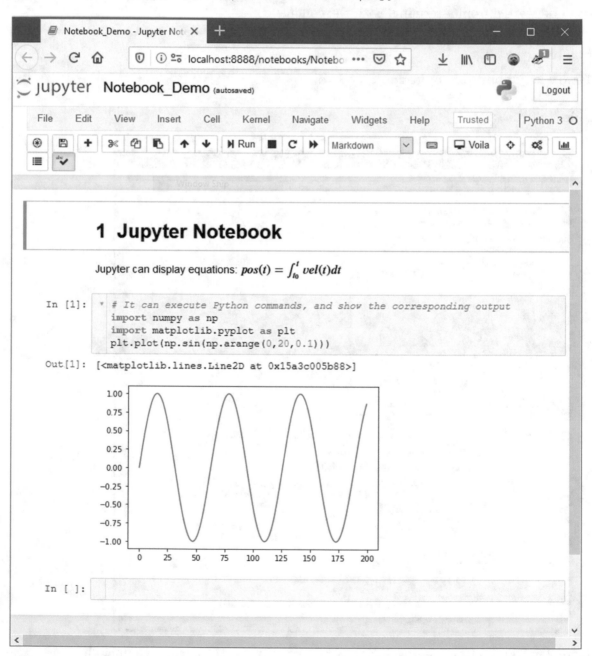

Figure 2.7: The *Jupyter Notebook* makes it easy to share research, formulas, and results.

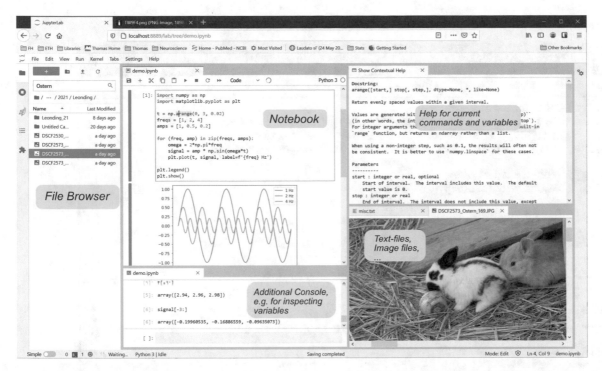

Figure 2.8: *JupyterLab* is the successor of the *Jupyter Notebook* and will eventually completely replace it. One can open additional consoles, e.g. for the inspection of variables, by right-clicking on the header-tab (here "demo.ipynb"). The "Contextual Help" can be opened with the shortcut CTRL+I. Individual panels can be arranged by clicking on the tab-header, and simply pulling them into the desirded position. JupyterLab can also displays text- files, image-files etc.

## JupyterLab

*Jupyterlab* is the successor to the *Jupyter Notebook*. As Fig. 2.8 shows, it extends the *Notebook* with very useful capabilities such as a file browser, easy access to commands and shortcuts, flexible image viewers etc. The file format stays the same as the *Notebook*, and both are saved as `.ipynb`-files.

### 2.3.2  First Session with the Qt Console

An important aspect of data analysis is interactive, visual inspection of the data. My personal preference is to start data analysis in the *Jupyter* Qt console.

I start my *IPython* sessions from the command-line, with the command `jupyter qtconsole`. (Under *WinPython*: if you have problems starting *Jupyter* from the `cmd` console, use the *WinPython Command Prompt* instead—it is nothing else but a command terminal with the environment variables set such that Python is readily found.)

To get started with Python and *IPython*, let me go step-by-step through the *IPython* session in Fig. 2.9:

- *IPython* starts out listing the version of *IPython* and Python that are used.

- **In [1]:** The first command imports the required Python packages. Note that by hitting `CTRL+Enter` one can execute multiline commands. (The command sequence gets executed after the next empty line.)

- **In [2]:** The command `t = np.arange(0,10,0.1)` generates a vector from 0 to 10, with a step size of 0.1. `arange` is a command in the *numpy* package.

- **In [3]:** Calculates `omega`. Note that the value of `pi` is only defined in *numpy*, and does not exist in Python!

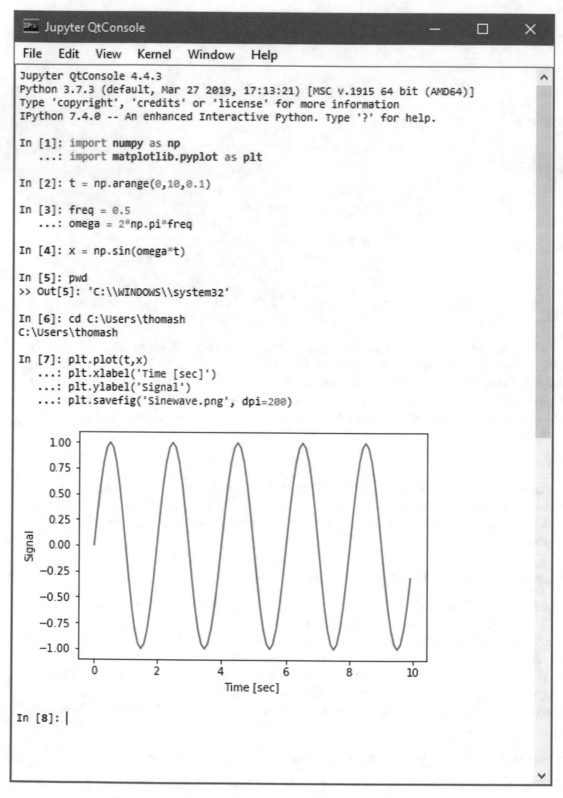

Figure 2.9: Sample session in the *Jupyter QtConsole*.

- **In [4]:** Since t is a vector, and sin is a function from *numpy*, the sine-value is calculated automatically for each value of t.

- **In [5]:** The IPython-"magic" command pwd stands for "print working directory", and does just that. Another "magic" command is cd. In Python scripts, changes of the current folder have to be performed with os.chdir(). However, tasks common with interactive computing, such as directory changes (%cd), bookmarks for directories (%bookmark), inspection of the workspace (%who and %whos), etc., are implemented as "IPython magic functions". If no Python variable with the same name exists, the "%" sign can be left away, as here.

- **In [7]:** *IPython* generates plots by default in the *Jupyter* Qt console, as shown in Fig. 2.9. Generating graphics files is also very simple: here I generate the PNG-file "Sinewave.png", with a resolution of 200 dots-per-inch.

I have mentioned above that *matplotlib* handles the graphics output. In *Jupyter* you can switch between inline graphs and output into an external graphics-window with %matplotlib inline and %matplotlib qt5 (see Fig. 2.10). (Depending on your version of Python, you may have to replace %matplotlib qt5 with %matplotlib or with %matplotlib tk.) An external graphics window allows zooming and panning, get the cursor position (which can help to find outliers), and get interactive input with the command plt.ginput. *matplotlib*'s plotting commands closely follow Matlab conventions.

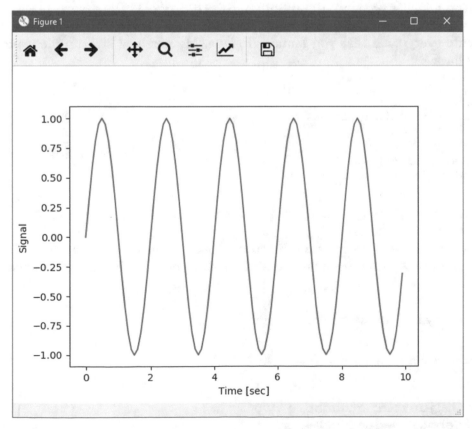

Figure 2.10: Graphical output window, using the Qt-framework. This allows you to pan (keyboard shortcut p), zoom (o), go home (h), or toggle the grid (g). With plt.ginput() one can also use it to get interactive input. (from Listing 2.4)

### 2.3.3   Personalizing IPython/Jupyter

When working on a new problem, I always start out with the *Qt-console* (see Fig. 2.6). Once the individual steps are working, one can use the *IPython* command `%history` to get the commands used. Then one can use either copy/paste, or save the history directly to a file with

```
%history -f [fname]
```

Then I switch to an IDE (integrated development environment), in my case *Wing*.

In the following, `<mydir>` has to be replaced with your home-directory (i.e. the directory that opens up when you run `cmd` in Windows, or `terminal` in Linux). And `<myname>` should be replaced by your name or your userID.

To start up *IPython* in a folder of your choice, and with personalized startup scripts, proceed as follows:

**In Windows**

- Type Win+R, and start a command shell with `cmd`

- In the newly created command shell, type `ipython profile create` . (This creates the directory `<mydir>\.ipython`).

- Add the Variable `IPYTHONDIR` to your environment (see above), and set it to `<mydir>\.ipython`. This directory contains the startup-commands for your *ipython*-sessions.

- Into the startup folder `<mydir>\.ipython\profile_default\startup` place a file for example with the name `00<myname>.py`, containing the startup commands that you want to execute every time that you launch *IPython*. My personal startup file contains the following lines:

```
import numpy as np
import matplotlib.pyplot as plt
import pandas as pd
import os
from scipy import stats
os.chdir(r'<working-dir>')
```

  This will import `numpy`, `matplotlib.pyplot`, and `pandas`, and then switches to the working directory of my choice.
  Note: since Windows uses \ to separate directories, but \ is also the escape character in strings, directory paths using a simple backslash have to be preceded by "r", indicating "raw strings".

- Generate a file "ipy.bat" in `<mydir>`, containing

```
jupyter qtconsole
```

- To customize the `jupyter qtconsole` type
  `jupyter notebook --generate-config`.
  This creates the file `jupyter_qtconsole_config.py` in your Jupyter folder. The Jupyter folder is in the sub-folder `.jupyter` in your home directory. In this file you find multiple options to configure your Qt Console, e.g. the distance between commands,

the editor used, the header displayed at the program start etc.
(The same procedure can be used to customize the `jupyter notebook` .)

To see all *Jupyter notebooks* that come with this book, for example, do the following:

- Type Win+R, and start a command shell with `cmd`

- Run the commands

  ```
  cd <ipynb-dir>
  jupyter lab
  ```

  where `<ipynb-dir>` is the directory where all the Jupyter noteboks are stored.

- Again, if you want, you can put this command sequence into a batch-file.

## In Linux

- Start a Linux terminal with the command `terminal`

- In the newly created command shell, execute the following command

  ```
  ipython
  ```

  (This generates a folder *.ipython*)

- Into the sub-folder `.ipython/profile_default/startup`, place a file with e.g. the name `00<myname>.py`, containing the lines

  ```
  import pandas as pd
  import os
  os.chdir(<mydir>)
  ```

- In your `.bashrc` file (which contains the startup commands for your shell-scripts), enter the lines

  ```
  alias ipy='jupyter qtconsole'
  IPYTHONDIR='~/.ipython'
  ```

- To see all *Jupyter* notebooks, do the following:

  - Go to `<mydir>`
  - Create the file `ipynb.sh`, containing the lines

    ```
    #!/bin/bash
    cd <ipynb-dir>
    jupyter lab
    ```

  - Make the file executable, with `chmod 755 ipynb.sh`

Now you can start "your" *Qt Console* by just typing `ipy`, and the *Jupyter Notebook* by typing `ipynb.sh`

**In Mac OS X**

- Start the *Terminal* either by manually opening *Spotlight* or the shortcut CMD + SPACE and entering Terminal and search for "Terminal".

- In *Terminal*, execute ipython, which will generate a folder under <mydir>/.ipython.

- Enter the command pwd into the *Terminal*. This lists <mydir>; copy this for later use.

- Now open *Anaconda* and launch an editor, e.g. *spyder-app* or TextEdit. Create a file containing the command lines you regularly use when writing code (you can always open this file and edit it). For starters you can create a file with the following command lines:

```
import pandas as pd
import os
os.chdir('<mydir>/.ipython/profile_<myname>')
```

- The next steps are somewhat tricky. *Mac OS X* by default hides the folders that start with "." (They can be shown with cmd-shift-.). So to access .ipython open File -> Save as .... Now open a *Finder* window, click the *Go* menu, select Go to Folder and enter <mydir>/.ipython/profile_default/startup. This will open a *Finder* window with a header named "startup". On the left of this text there should be a blue folder icon. Drag and drop the folder into the *Save as...* window open in the editor. IPython has a *README* file explaining the naming conventions. In our case the file must begin with 00-, so we could name it 00-<myname>.

- Open your .bash_profile (which contains the startup commands for your shell scripts), and enter the line
  alias ipy='jupyter qtconsole'

- To see all IPython Notebooks, do the following:

  - Go to <mydir>
  - Create the file ipynb.sh, containing the lines

    ```
    #!/bin/bash
    cd <ipynb-dir>
    jupyter lab
    ```

  - Make the file executable, with chmod 755 ipynb.sh

## 2.4 Workflow for Python Programming

### 2.4.1 Program Design

The best way to start a program is to take paper and pencil and to explicitly write down the algorithms to be implemented! This helps to clarify programming steps, which parameters have to be provided explicitly, and which have to be calculated during the execution of the program.

The example in Fig. 2.11 shows the first steps for a program that generates a sine-wave. Underlining the required parameters helps me to see which parameters need to be defined at the beginning of the program. And spelling out each step explicitly, e.g. the generation of a time-vector in line 4, clarifies which additional parameters arise in the program implementation.

This approach speeds up the implementation of a program and is an important first step in avoiding mistakes.

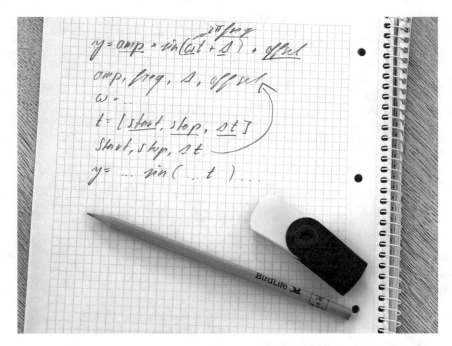

Figure 2.11: Still the best start to the successful development of a new program!

## 2.4.2  Converting Interactive Commands into a Python Program

*IPython* is very helpful in working out the command syntax and sequence. The next step is to turn these commands into a Python program with comments that can be run from the command-line. This section introduces a number of Python conventions and syntax features.

For me an efficient way to turn *IPython* commands into a script is to

- first obtain the command history with the command %hist or %history. (With the option -f you can save the history directly to the desired filename.)

- copy the history into a good IDE (integrated development environment): my preferred IDE is *Wing* (http://www.wingware.com/), because it provides a very comfortable and powerful working environment, with integrated code-versioning, testing tool, help-window etc., and with a powerful debugger (Fig. 2.12). The latest version of *spyder*, a free, science-oriented IDE that comes installed with *anaconda* and with *WinPython*, is also really impressive (*spyder4*, https://www.spyder-ide.org/). Other popular and powerful IDEs are *PyCharm* (https://www.jetbrains.com/pycharm/) and *Visual Studio Code* (https://code. visualstudio.com/).

- turn it into a working Python program by adding the relevant package information, substitute IPython magic commands, such as %cd, with their Python equivalent, and add more documentation.

Converting the commands from the interactive session in Fig. 2.9 into a program, we get

**Listing 2.4: python_script.py**

```python
1 """ Short demonstration of a Python script.
2 After a short one-line description of the content, the header can contain
3 further details.
4 """
5
6 # Import standard packages
7 import numpy as np
8 import matplotlib.pyplot as plt
9
10 # Generate the time-values
11 t = np.arange(0, 10, 0.1)
12
13 # Set the frequency, and calculate the sine-value
14 freq = 0.5
15 omega = 2 * np.pi * freq
16 x = np.sin(omega * t)
17
18 # Plot the data
19 plt.plot(t,x)
20
21 # Format the plot
22 plt.xlabel('Time[sec]')
23 plt.ylabel('Values')
24
25 # Generate a figure, one directory up, and let the user know about it
26 out_file = '../Sinewave.jpg'
27 plt.savefig(out_file, dpi=200, quality=90)
28 print(f'Image has been saved to {out_file}')
29
30 # Put it on the screen
31 plt.show()
```

The following modifications were made from the *IPython* history:

- The commands were put into a files with the extension ".py", a so called *Python module*.

- **1–4:** It is common style to precede a Python module with a header block. Multiline comments are given between triple quotes `"""` `<xxx>` `"""`. Below the first comment block describing the module there should be the information about author and date. (An excellent style-guide for Python can be found at https://pep8.org/.)

- **6:** Single-line comments use " # ".

- **7–8:** The required Python packages have to be imported explicitly. It is customary to import *numpy* as np, and *matplotlib.pyplot*, the *matplotlib* module containing all the plotting commands, as plt.

- **11 etc:** The numpy command `arange` has to be addressed through the corresponding package name, i.e. `np.arange`.

- **19 etc:** All the plotting commands are in the package `plt`.

- **26:** Care has to be taken with slashes in path names: in *Windows*, directories in pathnames are typically separated by `"\"`, which is also used as the escape-character in strings. To take `"\"` literally, a string has to be preceded by "r" (for "r"aw string), e.g. `r'C:\Users\Peter'` instead of `'C:\\Users \\Peter'`.

- **28:** f-strings were introduced in Python 3.6. With earlier versions, the corresponding syntax would be `print('Image has been saved to {0}'.format(out_file))`

- **31:** While *IPython* automatically shows graphical output, Python programs don't show the output until this is explicitly requested by `plt.show()`. The idea behind this is to optimize the program speed, only showing the graphical output when required. The output looks the same as in Fig. 2.11.

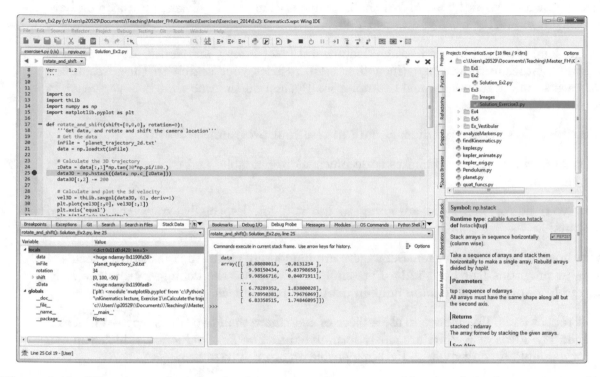

Figure 2.12: *Wing* is my favorite development environment, with one of the best existing debuggers for Python. **Tip:** If Python does not run right away in Wing, you may have to go to `Project -> Project Properties` and set the *Custom* `Python Executable` and/or `Python Path`.

## 2.5  Programming Tips

### 2.5.1  General Programming Tips

- Before you start programming, spell out the steps you have to do, and write them down as comments. A list of steps could look as follows:

  # Set the parameters

  # Select the input file

  # Read in the data

  # Filter the data

  # Show the results

  # Save the results to an outfile

  # Show the user the location of the outfile

  Not only does this help you to organize your code, it also provides a first rudimentary documentation of the program.

- Data analysis is an interactive task. Make use of the powerful interactive programming environment offered by *IPython/Jupyter*, and first develop your analysis step by step in a Qt console or in a *Jupyter* notebook.

- Once you have your data analysis—for the one block—going, grab the history with the command `history`, and turn it into a function. Think about what you want/need for the input, and what the output should be.

- And although I am repeating myself: Before you implement a mathematical algorithm, write it down on paper! This makes the implementation much quicker, because you have to spell out explicitly what you want to do.

- Use the help provided by the package documentations (*numpy*, *matplotlib*, and *scipy*) and by https://stackoverflow.com/. (In a first step, restrict your search to these resources: there are so many references and examples for Python on the web that it is very easy to get lost in them!)

- If possible, use some simple dummy data to test your program.

- Use clear variable names: it makes code much more readable, and easier to maintain in the long run.

- Know your editor well—you are going to use it a lot. Especially, know the keyboard shortcuts!

- Learn how to use the debugger. Debuggers are immensely useful to track down execution errors in programs (see Sect. 12.1). Personally, I always use the debugger from the IDE, and rarely resort to the *IPython* built-in debugger *pdb*, or *ipdb*.

- Don't repeat code. If you have to use a piece of code more than two times, write a function instead. The ideas of Python are nicely formulated in *The Zen of Python*, which you can see for example if you type in a Python console `import this`.

### 2.5.2 Python Tips

1. Have at least a brief look at the official Python style guide (https://pep8.org/), and stick to the standard conventions:

    - Every function should have a documentation string (in triple quotes """) on the line below the function definition.

    - Packages should be imported with their commonly used names:

    ```
    import numpy as np
    import matplotlib.pyplot as plt
    import scipy as sp
    import pandas as pd
    ```

2. To get the current directory, use `os.path.abspath(os.curdir)`. And in Python modules a change of directories can NOT be executed with `cd` (as in *IPython*), but instead requires the command `os.chdir(...)`.

3. Everything in Python is an object: to find out about "obj", use `type(obj)` and `dir(obj)`.

4. Use functions to avoid the duplication of code, and understand the `if __name__=='__main__':` construct (see p. 23).

5. If you have many of your personal functions in a directory `mydir` that is different from the current working directory, you can add that directory to your PYTHONPATH with the command

```
import sys
sys.path.append('mydir')
```

6. Avoid non-ASCII-characters, such as the German "ö, ä, ü, ß" or the French "é, è". Should you decide to use them anyway, you have to let Python know, by adding `# -*- coding: utf-8 -*-` in the first or second line of your Python module. This has to be done, even if the non-ASCII characters only appear in the comments! This requirement arises from the fact that Python will default to ASCII as standard encoding if no other encoding hints are given.

7. Make sure you know the basic Python syntax, especially the data structures. Try to use matrix multiplications instead of loops wherever possible: this makes the code nicer, and the programs much faster.

8. And along the same lines: note that many commands use an `axis` parameter, and can act on rows, columns, or on all data:

```
In [1]: mat = [[1, 2],
[3, 4]]

In [2]: np.max(mat)
Out[2]: 4

Tn [3]: np.max(mat, axis=0)
Out[3]: array([3, 4])

In [4]: np.max(mat, axis=1)
Out[4]: array([2, 4])
```

### 2.5.3 IPython/Jupyter Tips

1. Use *IPython* in the *Jupyter Qt Console* or the *Jupyter Notebook*, and customize your startup as described in Sect. 2.3.3: it will save you time in the long run!

2. For help on e.g. `plot`, use `help(plot)` or `plot?`. With one question mark the help gets displayed, with two question marks (e.g. `plot??`) also the source code is shown. Also check out the help tips shown with the command `%quickref`.

3. Use TAB-completion, for file- and directory-names, variable names, and for Python commands. This speeds up the coding, and helps to reduce typing mistakes.

4. To switch between inline and external graphs, use `%matplotlib inline` and `%matplotlib`.

5. You can use `edit <fileName>` to edit files in the local directory, and `%run <fileName>` to execute Python scripts in your current workspace.

6. The command `%bookmark` lets you quickly navigate to frequently used directories.

## 2.6  Python Alternatives

**Matlab** is a very popular program for data analysis. But with the sole exception of *Simulink*, Python can replace Matlab. And while Matlab licenses are very expensive, Python is completely free as in "Free beer!" A pretty fair comparison of the two can be found under https://pyzo.org/python_vs_matlab.html.

**R** is also a free language, and is primarily used for statistical data analysis and modeling (https://cran.r-project.org/)

## 2.7  Exercises

1. **Translating Data** Write a Python script that:

   - specifies two points, $P_0 = (0/0)$ and $P_1 = (2/1)$. Each point should be expressed as a Python list ([a,b]),
   - combines these two points to an np.array,
   - shifts those data, by adding *3* to the first coordinate, and *1* to the second,
   - plots a line from the original $P_0$ to the original $P_1$, and on the same plot also plot a line between the shifted values.

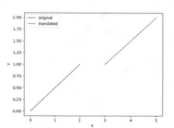

   More information on data display is presented in Chap. 4.

2. **Rotating a Vector** Write a Python script that specifies two points, $P_0 = (0/0)$ and $P_1 = (2/1)$.
   Then write a Python-function that:

   - takes a vector and an angle as input parameters,
   - rotates the vector by $25°$,
   - and returns the rotated vector.

   **Tip** A 2D rotation matrix is defined by

```
R = np.array([[np.cos(alpha), -np.sin(alpha)],
        [np.sin(alpha),  np.cos(alpha)]])
```

   If you want to experiment a bit with plots, you can try to

   - plot a green line from $P_0$ to $P_1$,
   - superpose this plot with a coordinate system, from –2 to +2,
   - superpose the rotated line in red, with increased line-thickness. (You can modify the width of a line with the plot parameter "linewidth=").

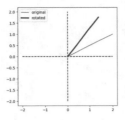

3. **Taylor Series**

   - Write a function that calculates the approximation to a sine and a cosine, to second order.
   - Write a script which plots the exact values, and superposes them with approximate values, in a range from –50 deg to +50 deg. (Command `plt.xlim`)
   - Save the resulting image to a PNG-file.

   **Tip** The second order approximations to sine and cosine are given by

$$\sin(\alpha) \approx \alpha$$
$$\cos(\alpha) \approx 1 - \frac{\alpha^2}{2}$$

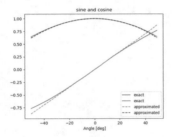

4. **First Steps with Pandas**

   - Generate a *pandas* DataFrame, with the x-column time stamps from 0 to 10 s, at a rate of 10 Hz, the y-column data values with a sine with 1.5 Hz, and the z-column the corresponding cosine values. Label the x-column "Time", the y-column "YVals", and the z-column "ZVals".
   - Show the head of this DataFrame.
   - Extract the data in rows 10–15 from "Yvals" and "ZVals", and write them to the file "out.txt".
   - Let the user know where the data have been written to.

# Reference

Scopatz, A., & Huff, K. (2015). *Effective computation in physics*. Sebastopol: O'Reilly Media.

# Chapter 3

# Data Input

This chapter shows how to read different types of data into Python. Thus it forms the link between the chapter on Python and the first chapter on statistical data analysis. It may be surprising, but reading data into the system in the correct format and checking for erroneous or missing entries is often one of the most time consuming parts of data analysis.

Data input can be complicated by a number of problems, like different separators between data entries (such as spaces and/or tabs), or empty lines at the end of the file. In addition, data may have been saved in different formats, such as *MS Excel*, *HDF5* (which also includes the Matlab-format), or in databases. This chapter gives an overview of where and how to start with data input.

## 3.1 Text

### 3.1.1 Visual Inspection

Reading in simple ASCII-text sounds like a trivial thing to do. A number of Python tools have been developed for data input. But regardless of which tool you use, you should always, check the following before trying to read in the data:

- Do the data have a header and/or a footer?

- Are there empty lines at the end of the file?

- Are there white-spaces before the first number, or at the end of each line? (The latter is a lot harder to see.)

- Are the data separated by tabs, and/or by spaces? (Tip: you should use a text-editor which can visualize tabs, spaces, and end-of-line (EOL) characters. Do *not* use *MS Excel* to inspect text files, since the representation of numbers in Excel depends on the "Region Settings" of your computer!)

- Are there missing values, and are they indicated consistently?

- Is the data type in each variable (column) consistent?

And after the data have been read in, check:

- Have the data in the first line been read in correctly?

- Have the data in the last line been read in correctly?

- Is the number of columns correct?

Figure 3.1 shows some of the variable aspects that commonly occur in text-files.

T. Haslwanter, *An Introduction to Hands-on Signal Analysis with Python*, https://doi.org/10.1007/978-3-030-57903-6_3

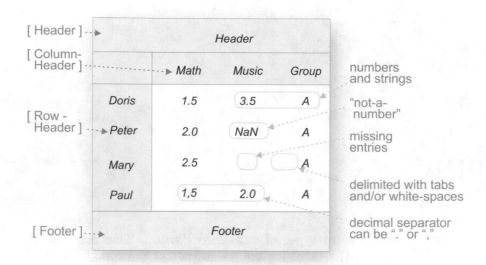

Figure 3.1: ASCII-files come in many varieties. **Left:** Labels in square brackets show optional data-file elements, corresponding to shaded data-file areas. **Right:** Importing text data can be encumbered by numerous tricky data features. Some of these pitfalls are indicated on the right.

### 3.1.2  Reading ASCII-Data

I strongly suggest that you start your data analysis by reading in and inspecting the data in the *Jupyter* QtConsole or in an *Jupyter* notebook. It allows you to move around much more easily, try things out, and quickly get feedback on how successful your commands have been. When you have found the correct command syntax to read in the data, you can obtain the command history with %history, copy it into your favorite IDE, and turn it into a program as described in the previous chapter.

While the *numpy* command np.loadtxt allows to read in simply formatted text data, most of the time it is easier to use the *pandas* command pd.read_csv as it provides significantly more powerful options for data-entry.

A typical workflow can contain the following steps:

- Changing to the folder where the data are stored.

- Listing the files in that folder.

- Selecting one of these files, and reading in the corresponding data.

- Checking if the data have been read in completely, and in the correct format.

These steps can be implemented in *IPython* for example with the following commands:

```
In [1]: import pandas as pd
In [2]: cd C:\Data\storage
In [3]: pwd        # Check if you were successful
In [4]: ls         # List the files in that directory
In [5]: in_file = 'data.txt'
In [6]: df = pd.read_csv(in_file)
In [7]: df.head()    # Check if first line is OK
In [8]: df.tail()    # Check the last line
```

After "In [6]" I often have to adjust the options of pd.read_csv to read in the data correctly. Make sure to check that the number of column headers is equal to the number of columns that you expect. It can happen that everything gets read in—but into one large single column!

**Simple Text-Files**

For example, a file data.txt with the following content

```
1, 1.3, 0.6
2, 2.1, 0.7
3, 4.8, 0.8
4, 3.3, 0.9
```

can be read in and displayed with

```
In [9]: data = np.loadtxt('data.txt', delimiter=',')

In [10]: data
Out[10]:
    array([[ 1. ,   1.3,   0.6],
           [ 2. ,   2.1,   0.7],
           [ 3. ,   4.8,   0.8],
           [ 4. ,   3.3,   0.9]])
```

where data is a *numpy* array. Without the flag delimiter=',', the function np.loadtxt crashes. An alternative way to read in these data is with

```
In [11]: df = pd.read_csv('data.txt', header=None)

In [12]: df
Out[12]:
     0    1    2
0    1  1.3  0.6
1    2  2.1  0.7
2    3  4.8  0.8
3    4  3.3  0.9
```

where df is a *pandas* DataFrame. Without the flag header=None, the entries of the first row are falsely interpreted as the column labels as shown in the next step:

```
In [13]: df = pd.read_csv('data.txt')   # Warning: incorrect result!

In [14]: df
Out[14]:
        1    1.3    0.6
    0   2    2.1    0.7
    1   3    4.8    0.8
    2   4    3.3    0.9
```

Note that the *pandas* function `pd.read_csv` already recognizes the first column as integer, whereas the second and third columns are correctly identified as float.

### More Complex Text-Files

The advantage of using *pandas* for data input becomes clear with more complex files. Take for example the input file "data2.txt", containing the following lines (including the footer):

```
ID, Weight, Value
1, 1.3, 0.6
2, 2.1, 0.7
3, 4.8, 0.8
4, 3.3, 0.9

Those are dummy values, created by ThH.
May, 2020
```

One of the input flags of `pd.read_csv` is `skipfooter`, so we can read in the data easily with

```
In [15]: df2 = pd.read_csv('data.txt',
         skipfooter=3,
         delimiter='[ ,]*')
```

The last option, `delimiter='[ ,]*'`, is a *regular expression* (see below) specifying that "one or more spaces and/or commas may be used to separate entry values". Also, when the input file includes a header row with the column names, the data can be accessed immediately with their corresponding column name, e.g.:

```
In [16]: df2
Out[16]:
   ID   Weight   Value
0   1      1.3     0.6
1   2      2.1     0.7
2   3      4.8     0.8
3   4      3.3     0.9

In [17]: df2.Value
Out[17]:
0    0.6
1    0.7
2    0.8
3    0.9
Name: Value, dtype: float64
```

**Tip:** an option of the command `pd.read_csv` that is frequently useful is `delim_white-space`. When this parameter is set to `True`, one or more white-spaces (spaces or tabs) are taken as a single separator.

### 3.1.3  Regular Expressions

Working with text data often requires the use of simple *regular expressions*. Regular expressions are a very powerful way of finding and/or manipulating text strings, and their syntax is independent of the programming language used. They are directly supported by most programming languages. Many books have been written about them, and good, concise information on regular expressions can be found on the web, for example at:

- https://www.debuggex.com/cheatsheet/regex/python provides a convenient cheat sheet for regular expressions in Python.

- http://www.regular-expressions.info gives a comprehensive description of regular expressions.

Below you find two examples how *pandas* can make use of regular expressions:

1. Reading in data from a file, separated by a combination of commas, semicolons, or whitespaces:

```
df = pd.read_csv(inFile, sep='[ ;,]*')
```

The square brackets ("[...]") indication a *combination* the elements inside the brackets. And the star ("*") indicates *one or more* of the preceding element.

2. Extracting columns with certain name-patterns from a *pandas* DataFrame. In the following example, all columns starting with Vel are extracted and combined:

```
In [18]: data = np.round(np.random.randn(100,7), 2)

In [19]: df = pd.DataFrame(data, columns=['Time',
            'PosX', 'PosY', 'PosZ', 'VelX', 'VelY', 'VelZ'])

In [20]: df.head()
Out[20]:
    Time  PosX  PosY  PosZ  VelX  VelY  VelZ
0   0.30 -0.13  1.42  0.45  0.42 -0.64 -0.86
1   0.17  1.36 -0.92 -1.81 -0.45 -1.00 -0.19
2  -3.03 -0.55  1.82  0.28  0.29  0.44  1.89
3  -1.06 -0.94 -0.95  0.77 -0.10 -1.58  1.50
4   0.74 -1.81  1.23  1.82  0.45 -0.16  0.12

In [21]: vel = df.filter(regex='Vel*')

In [22]: vel.head()
Out[22]:
    VelX  VelY  VelZ
0   0.42 -0.64 -0.86
1  -0.45 -1.00 -0.19
2   0.29  0.44  1.89
3  -0.10 -1.58  1.50
4   0.45 -0.16  0.12
```

## 3.2 Excel

There are two approaches to reading a *MS Excel* file in *pandas*: the function read_excel, and the class ExcelFile.[1]

- read_excel is for reading one file with file-specific arguments (i.e. identical data formats across sheets).

- ExcelFile is for reading one file with sheet-specific arguments (i.e. different data formats across sheets).

---

[1]The following section has been taken from the *pandas* documentation.

Choosing the approach is largely a question of code readability and execution speed.

The following commands show equivalent class and function approaches to read a single sheet:

```
# using the ExcelFile class
xls = pd.ExcelFile('path_to_file.xls')
data = xls.parse('Sheet1', index_col=None,
                    na_values=['NA'])

# using the read_excel function
data = read_excel('path_to_file.xls', 'Sheet1',
        index_col=None, na_values=['NA'])
```

If this fails, give it a try with the Python package *xlrd*.

## 3.3  Matlab

The best input solution for *Matlab* files depends on the complexity of the files. For .mat files which contain only strings, numbers, vectors, and matrices, the easiest solution is scipy.io.loadmat.

The following commands return a string, a number, a vector, and a matrix variable from a Matlab file data.mat.

```
from scipy.io import loadmat
data = loadmat(matlab_file, squeeze_me=True)

# Field names are the names of the Matlab variables
text = data['my_text']
number = data['float_number']
vector = data['my_vector']
matrix = data['my_matrix']
```

If the .mat file also contains cells and structures, but no more complex data structures (e.g. arrays with more than 2 dimensions or with complex numbers, sparse arrays etc.), then the package mat4py is great. Note that mat4py is typically not included in the common Python distributions, and thus has to be installed by hand (pip install mat4py). It returns the data in simple Python datatypes (specifically it returns arrays as list and not as np.array).

```
import mat4py

data = mat4py.loadmat(matlab_file)
array_data = np.array( data['my_matrix'] )
cell = data['my_cell']
```

python [2] matlab_data.py shows different ways to read in data .mat-files.

## 3.4  Binary Data

If space and/or bandwidth is at a premium it may be desirable to save data in binary format. In that case more care is required, since binary representations offer many different options. For signal processing, two useful options are the .npz file format provided by *numpy*, or so-called

[2]https://github.com/thomas-haslwanter/sapy/blob/master/src/code_quantlets/matlab_data.py

"structured arrays". The former one is useful if the data are read in by Python programs again. The latter one is preferable if the data should be further processed by other applications.

### 3.4.1 NPZ Format

The first example shows how to save data to `.npz` format, which is a zipped archive of files named after the variables they contain. The archive is not compressed and each file in the archive contains one variable in .npy format.

```
In [1]: import numpy as np
In [3]: t = np.arange(0, 10, 0.1)                    # Generate np arrays
In [4]: x = np.sin(t)
In [5]: data_dict = {'time': t, 'signal': x}         # Group them in a dictionary
In [6]: out_file = 'binary'
In [7]: np.savez(out_file, **data_dict)              # Save dictionary to '.npz'-file
In [8]: new_data = np.load(out_file + '.npz')        # Load '.npz'-file
In [9]: plt.plot(new_data['time'], new_data['signal']) # Use loaded data
In [10]: new_dict = dict(new_data)                   # Convert them to dictionary
```

### 3.4.2 Structured Arrays

A good concise introduction to working with binary data in Python is https://www.devdungeon.com/content/working-binary-data-python#bytesio

To following commands give a taste of binary data representation and storage:

```
In [11]: import struct
         import numpy as np
In [12]: one = struct.pack('hhf', 1, 2, np.pi)
In [13]: one

Out[13]: b'\x01\x00\x02\x00\xdb\x0fI@'

In [14]: two = struct.pack('hhf', 3, 4, 2*np.pi)
In [15]: two

Out[15]: b'\x03\x00\x04\x00\xdb\x0f\xc9@'

In [16]: data_file = 'data.raw'
In [17]: with open(data_file, 'wb') as fh:
             fh.write(one)
             fh.write(two)
In [18]: ls data.raw
    [...]
    22.05.2020  17:21                16 data.raw
    1 File(s)            16 bytes
    [...]

In  [19]: dt = np.dtype([('i1', np.short), ('i2', np.short),
                        ('float_val', np.float32)])
In [20]: data = np.fromfile(data_file, dtype=dt)
In [21]: data

Out[21]: array([((1, 2, 3.1415927), (3, 4, 6.2831855)],
    dtype=[('i1', '<i2'), ('i2', '<i2'), ('float_val', '<f4')])
```

**Line 12** The command `struct.pack` returns a bytes object containing here the values *1,* *2, pi,* packed according to the format string 'hhf'. That string indicates that the two integers are stored as `short` ('h'), and the value `np.pi` as `float` ('f').

**Line 17** The `with`-statement is a convenient shortcut for file-IO. It ensures that the file gets correctly opened at the beginning of the block, and closed at the end.

**Line 18** Shows that the output file has exactly 16 byte (*2\*4* byte for the short integer, and *1\*8* byte for the float).

**Line 20** `np.fromfile` returns what is called a "structured array", and can—in contrast to standard *numpy* arrays—contain different datatypes.

**Line 21** : The read in data correspond to the ones defined in lines 2 & 4. The second line in the output indicates the data types: for example `'<i2'` indicates a 2-byte integer ("i2"), represented in little-endian byte-order ("i").

If all binary data in the structured array have the same format, they can be converted to a 2-dimensional array with `array_2d = np.array(structured_array.tolist())`

## 3.5   Images

Surprisingly, reading in image files is typically a lot simpler than reading in text-files! Common image files can be read in with `plt.imread`. For color images with $m$ rows and $n$ columns, the resulting data typically form an $m \times n \times 3$-matrix, where the third dimension contains the RGB (*"R"ed, "G"reen, "B"lue*) image planes (see also Fig. 5.27). The script below shows how to read in an image, manually select points, and mark those points on the image (Fig. 3.2).

**Listing 3.1: mark_planets.py**

```python
"""Show how to read in images, and mark selected locations"""

import numpy as np
import matplotlib.pyplot as plt

# Set the filenames
in_file = r'..\..\Data\Saturn.jpg'
out_file = 'Saturn_marked.jpg'

# Get the data
img = plt.imread(in_file)

# Select the planets
plt.imshow(img)

fig = plt.gcf()
fig.canvas.set_window_title('Please select the moons:')

sel_pts = plt.ginput(4)
#sel_pts = np.array(selection, dtype='uint16')

# Mark the planets
ax = plt.gca()
for ii in range(len(sel_pts)):
    ax.add_artist(plt.Circle(sel_pts[ii],
        radius=30,
        color='g',
        lw = 4,
     fill=False))

# Show the result
plt.pause(3)
```

Figure 3.2: Color image of Saturn, with four moons manually selected and indicated (from Listing 3.1).

```
# Save the resulting file
plt.savefig(out_file, dpi=200, quality=90)
print(f'Image saved to {out_file}')

plt.show()
```

## 3.6   Videos

For working with videos *OpenCV* is the package of choice (https://opencv.org/). Since the Python wrapper is closely based on the original C++ implementation, its syntax is less "pythonic" than other packages. Figure 3.3 contains a few screenshots from the movie played by

Figure 3.3: OpenCV is the ideal tool for working with videos (from Listing 3.2).

```
Listing 3.2: openCV.py
```

```python
""" Show how to work with video-data using OpenCV

Note that the package OpenCV has to be installed for this program to work.
"""

# Import the required packages
import numpy as np
import matplotlib.pyplot as plt
import cv2

# Set the video file, and open it
video_file = r'..\Data\Pleasure.mp4'
cap = cv2.VideoCapture(video_file)

# Set the parameters
out_base =  r'..\Data\Pleasure_out'
dt = 25 # [msec]
counter = 0

# Show the movie
while(cap.isOpened()):
    ret, frame = cap.read()        # Get the next frame

    if ret == True:
        # Convert to gray
        gray = cv2.cvtColor(frame, cv2.COLOR_BGR2GRAY)

        # Show the frame for 'dt' msec, or until the user hits 'q'
        cv2.imshow('frame',gray)
        if cv2.waitKey(dt) & 0xFF == ord('q'):
            break

        # Save every 50-th frame
        if not np.remainder(counter, 50):
            out_file = out_base + str(counter) + '.jpg'
            plt.imshow(gray, cmap='gray')
            plt.savefig(out_file)
            plt.close()
            print(f'{out_file} saved')

        counter += 1

    else:
        # Return after the end of the movie
        break

# Clean things up
cap.release()
cv2.destroyAllWindows()
```

## 3.7   Sound

Working with sound is made a bit more challenging by the multitude of sound-formats that
are frequently used. For wav-files, the scipy module `scipy.io.wavfile` provides functions
for reading and writing sound. When working with other formats, the easiest way is to install `ffmpeg`, the open-source "Swiss-army-knife" software for working with video and sound.
(`ffmpeg` can be obtained from http://ffmpeg.org.) In order to make it simple to read, write,
and play sound files with arbitrary format, I have written the package `scikit-sound`. It can
be installed from the command-line with

```
pip install scikit-sound
```

and the documentation is available under http://work.thaslwanter.at/sksound/html/. Using these two applications, Neil Armstrong's famous words when he first stepped out on the moon can for example be loaded and played with

```
in_file = r'.\data\a11step.wav'
mySound = sounds.Sound(in_file)
mySound.play()
```

## 3.8 Zipped Archives on the WWW

 python **Code:** readZip.py (Appendix A) shows how to directly read in data from an *Excel* file which is stored in a zipped archive on the web.

## 3.9 Other Formats

**Clipboard** data from the system clipboard can be import directly with

```
pd.read_clipboard()
```

**Other file formats** Also SQL databases and a number of additional formats are supported by *pandas*. The simplest way to access them is typing pd.read_ + TAB, which shows all currently available options for reading data into *pandas* DataFrames.

## 3.10 Exercises

1. **Reading in Data** Read the data from the following files into your workspace. (If the files are not yet available, they can be generated by running the script S3_data_gen.py.)

   **data.csv** Comma-separated data file

   **data_tab.txt** Tab-separated data file

   **data_modified.txt** Tab-separated data file, with header

   **data.xls** Excel file

   **data.mat** Matlab file

2. **Modifying Text Files: Imaginary Numbers (hard)** The file .\data\imaginary.txt contains the real- and the imaginary-part of complex numbers, including a header for each column. Write a Python script that reads in those data, and adds the polar representation of each data-point (radius/angle [rad]) to each line, with the name of the out-file imaginary_out.txt.
   **Tips**

   - For the resulting out-file, separate header labels and numbers by a simple tab.

3. **Mixed Inputs**

   - The file (.\data\data\swim100m.csv) contains values and strings. Read in the data, and show the first 5 and the last 5 data points.
   - The MS-Excel file (.\data\data_others\Table 2.8 Waist loss.xls) contains some data lower down in the file. Read in the data, and show the last 5 data points.

**Hard:** Read in the same file, but this time from the zipped archive https://work.thaslwanter.
at/sapy/GLM.dobson.data.zip.

4. **Binary Data** The file `.\data\data.raw` has a 256 byte header, followed by triplets of
   data (`(t, x, y)`) stored in `float` representation. Read in those data, and plot x and
   y versus t.
   Tips:

 python **Code:** How to write data in different formats, and how to produce formatted
text strings, is shown in `S3_data_gen.py` (Sect. B.3). That script also produces the input files
for the exercises for this chapter.

# Chapter 4

# Data Display

This chapter presents the basic concepts of plotting in Python. It also provides help in turning Python plots into good looking figures for presentations. Examples of different 2D and 3D plot types provide a first look into the capabilities of *matplotlib*, the dominant plotting package.

We will start out with the movement of a particle along a simple trajectory, showing position and velocity of a particle that moves along a figure-eight path (see Fig. 4.1) as a function of time.

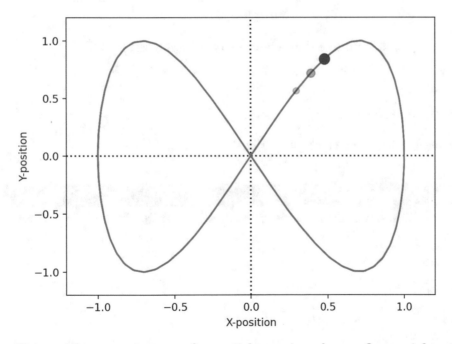

Figure 4.1: x-y trajectory of a particle moving along a figure eight.

Based on the position and velocity of such a particle, the next section then explains the basic concepts for generating plots in Python. It also presents code-samples for a number of helpful features, such as positioning figures on the computer screen, or querying keyboard input for figures. The last section describes how Python figures can be exported to some common, vector-based graphics programs, to facilitate the preparation of figures for different types of presentation formats.

© The Author(s), under exclusive license to Springer Nature Switzerland AG 2021
T. Haslwanter, *An Introduction to Hands-on Signal Analysis with Python*,
https://doi.org/10.1007/978-3-030-57903-6_4

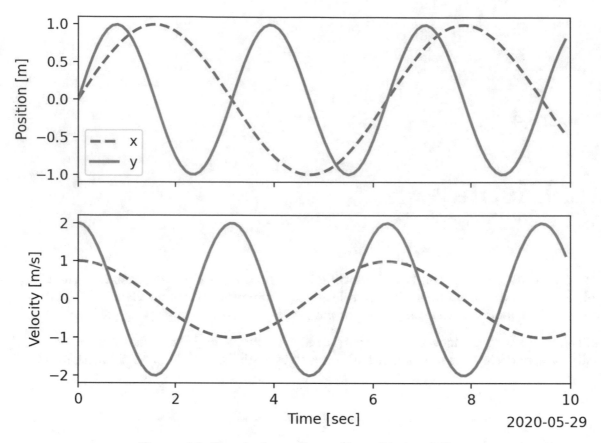

Figure 4.2: Simple demo figure (from Listing 4.1).

## 4.1   Introductory Example

The code in Listing 4.1 produces the plot in Fig. 4.2.

Listing 4.1: simple_figure.py

```
""" Basic plotting commands, by showing position and velocity of two curves """

# Import standard packages
import numpy as np
import matplotlib.pyplot as plt
from datetime import date

# Generate the data
t = np.arange(0,10,0.1)

x = np.sin(t)          # position
y = np.sin(2*t)        # velocity

vx = np.cos(t)
vy = 2*np.cos(2*t)


# Put the position in the top plot, and the velocity in the bottom
# Ensure that when you zoom into one, the other also gets adjusted
fig, axs = plt.subplots(nrows=2, ncols=1, sharex=True)

# Here a plot-parameter is modified during the creation of the figure element.
# The parameters 'linewidth=' and 'lw=' are equivalent
axs[0].plot(t, np.column_stack([x,y]), linewidth=2)
```

```
axs[1].plot(t, np.column_stack([vx,vy]), lw=2)

# Add the axis labels
axs[0].set_ylabel('Position [m]')
axs[1].set_xlabel('Time [sec]')
axs[1].set_ylabel('Velocity [m/s]')

# Set the x-limit (Note that since the x-axes are shared, we only have to do
    this once!)
axs[0].set_xlim([0, 10])

# Also put the date on the figure
fig.text(0.8, 0.02, date.isoformat(date.today()))

# Properties of figure elements can also be changed after they have been drawn:
for ax in axs:
    lines = ax.get_lines()
    lines[0].set_linestyle('dashed')

# Add a legend to the first axis
axs[0].legend(['x', 'y'])

# Save the figure
out_file = 'simple_figure.jpg'
plt.savefig(out_file, dpi=200, quality=90)

# Always inform the user if any file has been added or modified on the computer
print(f'Image saved to {out_file}')

plt.show()
```

The following basic rules can help to optimize figures:

- Illustrations should contain as few elements as possible, but as many as necessary (Fig. 4.2).

- Always label the axes, and include axes units in the labels!!

- Minimize the amount of reading the user has to do. For example

    - If two axes are above each other with the same x-scale, only use x-tick-labels for the lower axis.
    - If two axes have the same line style and labels, only use one legend.

- Consider putting a date on your figures. This may help you later on to interpret them—especially if you modify some parameters in data analysis.

- Separate the generation of the figure elements from the formatting (see next section). This helps to clarify the code.

- When saving a figure to a file, always tell the user where a file has been generated/modified, and the name of that file!

## 4.2   Plotting in Python

The first step in data analysis should always be a visual inspection of the raw-data (Fig. 4.3).

The dominant task of the human cortex is to extract visual information from the activity patterns on the retina. Our visual system is therefore exceedingly good at detecting patterns in visualized data sets. As a result, one can almost always *see* what is happening before it

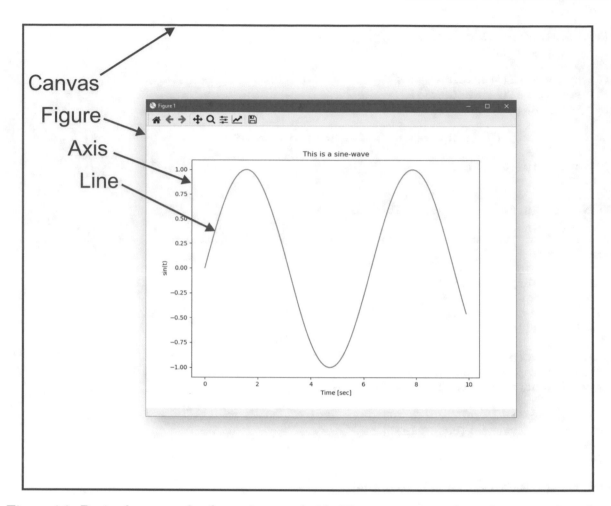

Figure 4.3: Basic elements of a figure in *matplotlib*. The `canvas` can be a figure window, but it can also be a PDF-file, or a section in a browser. It only needs to be addressed rarely, for example to position the figure window on the computer screen.

can be demonstrated through quantitative analysis of the data. Visual data displays are also helpful at finding extreme data values, which are often caused by mistakes in the execution of the paradigm or mistakes in the data acquisition.

In practice the display of data can be tricky as there are so many options: graphical output can be displayed as a picture in an HTML-page or in an interactive graphics window, plots can demand the attention of the user (so called "blocking figures") or can automatically close after a few seconds, etc. This section will therefore focus on general aspects of plotting data; the next section will then present different types of plots, e.g. histograms, error bars and 3D-plots.

The Python core does not include any tools to generate plots. This functionality is added by other packages. By far the most common package for plotting is *matplotlib*. If you have installed Python with a scientific distribution like *WinPython* or *Anaconda*, it will already be included. *Matplotlib* is intended to mimic the style of Matlab. As such, users can either generate plots in a functional style ("Matlab style"), or in the traditional, object-oriented Python style (see below).

*Matplotlib* (https://matplotlib.org/) contains different modules and features:

**matplotlib.pyplot** This is the module that is commonly used to generate plots. It provides the interface to the plotting library in *matplotlib*, and is by convention imported in Python functions and modules with

```
import matplotlib.pyplot as plt.
```

*pyplot* handles lots of little details, such as creating figures and axes for the plot, so that the user can concentrate on data analysis.

**Matplotlib.mlab** Contains a number of functions that are commonly used in Matlab, such as `find`, `griddata`, etc.

**backends** *Matplotlib* can produce output in many different formats, which are referred to as "backends":

- In a *Jupyter* notebook or in a *Jupyter* Qt console, the command `%matplotlib inline` directs output into the current browser window.

- In the same environment, `%matplotlib qt5`[1] directs the output into a separate graphics window (Fig. 2.10). This allows panning and zooming the plot, and interactive selection of points on the plot by the user with the command `plt.ginput`.

- With `plt.savefig` output can be easily directed to external files, e.g. in PDF, PNG, or JPG format.

*pylab* is a convenience module that bulk imports `matplotlib.pyplot` (for plotting) and `numpy` (for mathematics and working with arrays) in a single name space. Although many examples use *pylab*, it is no longer recommended.

The easiest way to find an implementation of one of the many image types that *matplotlib* offers is to browse the *matplotlib* gallery (https://matplotlib.org/gallery/), and copy the corresponding Python code into your program.

Other interesting plotting packages and resources are:

- Also *pandas* (which builds on *matplotlib*) offers many convenient ways to visualize DataFrames, because the axis- and line-labels are already defined by the column names.

  https://pandas.pydata.org/pandas-docs/stable/user_guide/visualization.html

- *PyQtGraph* is a powerful library for scientific graphics and graphical user interfaces (GUIs). The library is very fast due to its heavy leverage of numpy for number crunching and Qt's GraphicsView framework for fast display. (See also Sect. 12.4.2.) (http://pyqtgraph.org/)

- *bokeh* is a Python interactive visualization library that targets modern web browsers for presentation. *bokeh* allows the creation of interactive plots, dashboards, and data applications (see Fig. 4.4). (https://docs.bokeh.org/)

- *Dash* from https://plotly.com is used for the generation of web-based analytics apps.

- *ggplot* for Python. It emulates the *R*-package *ggplot* which is loved by many *R*-users. (https://github.com/yhat/ggpy)

## 4.2.1 Functional and Object-Oriented Approaches

Python plots can be generated in a functional, Matlab-like style, or in an object oriented, more pythonic way. These styles are all perfectly valid, and each have their pros and cons. The only caveat is to avoid mixing the coding styles in your own code.

First, consider the frequently used functional "pyplot" style:

---

[1] Depending on your build of Python, this command may also be `%matplotlib` or `%matplotlib tk`.

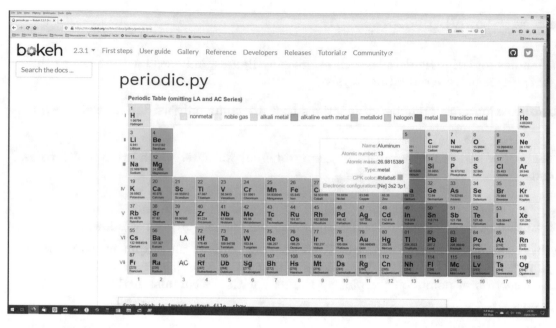

Figure 4.4: Bokeh makes it easy to generate interactive graphics. Here information about the element currently under the cursor is displayed.

```
# Import the required packages, with their conventional names
import matplotlib.pyplot as plt
import numpy as np

# Generate the data
x = np.arange(0, 10, 0.2)
y = np.sin(x)

# Generate the plot
plt.plot(x, y)

# Display the plot on the screen
plt.show()
```

Note that the creation of the required figure and axis is done automatically by *pyplot*.

Second, a more pythonic, object oriented style, which may be clearer when working with multiple figures and axes. Compared to the example above only the section entitled *"# Generate the plot"* changes:

```
# Generate the plot
fig = plt.figure()            # Generate the figure
ax = fig.add_subplot(111)     # Add an axis to that figure
ax.plot(x,y)                  # Add a plot to that axis
```

So, why all the extra typing as one moves away from the pure Matlab-style? For very simple plots like this example, the only advantage is academic: the wordier styles are more explicit, and clearer as to where things come from and what is going on. For more complicated applications, this explicitness and clarity becomes increasingly valuable, and the richer and more complete object-oriented interface will likely make the program easier to write and to maintain. For example, the following lines of code produce a figure with two plots above each other, and clearly indicate which plot goes into which axis:

```
# Import the required packages
import matplotlib.pyplot as plt
```

```python
import numpy as np

# Generate the data
x = np.arange(0, 10, 0.2)
y = np.sin(x)
z = np.cos(x)

# Generate the figure and the axes
fig, axs = plt.subplots(nrows=2, ncols=1)

# On the first axis, plot the sine and label the ordinate
axs[0].plot(x,y)
axs[0].set_ylabel('Sine')

# On the second axis, plot the cosine
axs[1].plot(x,z)
axs[1].set_ylabel('Cosine')

# Display the resulting plot
plt.show()
```

python **Code**[2]: "getting_started.py" gives a short demonstration of Python for scientific data analysis.

### 4.2.2   Interactive Plots

*Matplotlib* provides different ways to interact with the user. The examples below may help to quickly create user interfaces for interactive visual data inspection. They show how to

- position figures on the screen

- pause between two plots, and proceed automatically after a few seconds

- proceed on a click or keyboard hit

- evaluate keyboard entries

- interactively show information about individual data points, helping to find and evaluate outliers.

python **Code:** `interactive_plots.py` (Appendix A) shows the corresponding source code.

## 4.3   Saving a Figure

*Matplotlib* offers multiple options for the export of figures. In general one distinguishes between pixel-based options (such as PNG or TIFF), vector-based options (such as SVG), or compressed formats (such as JPEG). Three of those formats are particularly useful:

**JPEG** ("Joint Photographic Experts Group") is a compressed format, which is recommendable for figures that are going to be printed or imported in other documents. When exporting to JPEG, note that it is recommendable to specify a relatively high quality rate in order to avoid compression artifacts.

**SVG** ("Scalable Vector Graphic") is useful when figures are to be modified in external applications. SVG is a vector graphic format, and allows you to modify different aspects of the data (LineStyle, LineWidth, etc) in an external application as demonstrated in the next section.

---

[2]https://github.com/thomas-haslwanter/sapy/blob/master/src/code_quantlets/getting_started.py.

**PNG** ("Portable Network Graphics") is a compressed raster graphic format. It is commonly used on the Web.

```
out_file = 'my_figure.jpg';
# specify a resolution of 200 dots-per-inch, and 90% quality
plt.savefig(out_file, dpi=200, quality=90);

print(f'The figure has been saved to {out_file}')
```

**Tip:** One should always let the user know when the program generates a new file, or modifies an existing one.

## 4.4   Preparing Figures for Presentation

### 4.4.1   General Considerations

All elements in Python, and therefore all elements of a *graphical user interface (GUI)*, are objects (see Fig. 4.3). As a result characteristics of figure elements can be determined or modified at three stages:

1. at the creation of the original figure elements,

2. after the creation by modifying the properties of the element (for example, a line has a `linewidth`, a `color`, etc.),

3. or in an external graphics program.

I strongly recommend doing only the basic figure adjustments in Python. Then save the figure in SVG-format so that you can modify the individual elements and all the fine details, such as line-width and -color, text-size, etc in an external vector graphics program. Properties such as text-size may change depending on how the figure is going to be used, and it is typically much easier to modify these properties in a graphics program instead of having to go back and also redo the computation of the figure. (This requires that data analysis program is still working, that the input data are still available, that and that they are also in the correct location, etc.)

**Popular Vector Graphics Programs**

**Adobe Illustrator** I used to be an Adobe fan. But since Adobe has eliminated the option to buy software and only offers software subscriptions I have turned my back on it.

**Affinity Designer** The new kid on the block, affordable, and getting very good reviews.

**CorelDraw** Commercial, cheaper than Illustrator, but more powerful than Inkscape.

**Inkscape** An open source, free vector drawing program.

**Type of Figures**

While the specific steps may depend slightly on the program used, the overall procedure should be almost the same.

- The SVG-file created in Python has to be imported into the vector graphics program.

- All elements have to be ungrouped. (Note that there may be grouped elements within other groups!)

- Now the individual elements can be modified.

- For clarity, it is often useful to use *layers* for the different elements of the figure (lines, labels, annotations).

### 4.4.2 Modifying SVG Figures

**CorelDraw**

In the following the workflow is described for modifying the figure obtained from `sine.svg`, a simple sine wave generated with *matplotlib*, with *CorelDraw*:

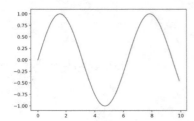 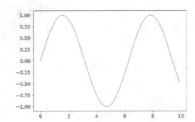 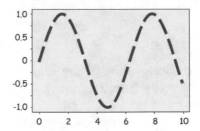

Figure 4.5: **Left:** original SVG figure. **Center:** *Wireframe* view of the figure imported into CorelDraw. Note the background frame around the figure, which is not visible in the standard view! **Right:** figure, with the line-style, background, and the tick-labeling modified, and exported to a JPEG-file.

- `File | Import` → select `sine.svg` (Fig. 4.5, left)

- Specify the area where you want to insert the figure

- `Object | Group | Ungroup All Objects`

- Select and remove the background. (This may have to be done twice, as there may be two background layers.) Note that the background is often white, and only becomes visible in the *wireframe* view of the illustration (Fig. 4.5, center)

- Select the line

- Change line-color (by right clicking the desired color) and fill-color (left-clicking) of your elements. ("No color" can be selected by clicking the color ⊠)

- Adjust the line-width and the line-style

For adjusting text elements and labels I typically proceed by adding additional "Layers", and moving all the text (axis labels, line labels, etc) into that layer. This way I can "lock" the figure and adjust the labeling without having to worry about the curves (Fig. 4.6).

- Open the `Object Manager` ( `Windows | Dockers | Object Manager`)

- Click the black arrow/triangle in the upper-right corner, or the "New Layer" symbol in the lower left corner of the `Object Manager`, and add a `New Layer`

- Call the original layer "Figure", and the new layer "Labels"

- Select all the Text elements, and move/drag them from "Figure" to "Labels"

- Lock "Figure"

- Select "Labels", and make all the desired text adjustments.

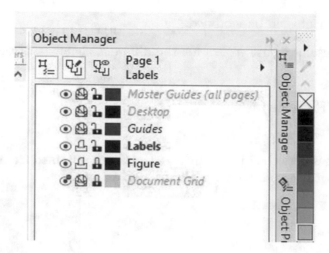

Figure 4.6: Screenshot of the *Object Manger* in *CorelDraw*. The black arrow in the top-right corner lets you generate a new layer.

**Affinity Designer**

- File | Place → select sine.svg

- Specify the area where you want to insert the figure

- Double-click the figure

- Layer | Ungroup All

- Select and remove the background. (This may have to be done twice, as there may be two background layers.) Note that the background is often white, and only becomes visible in the "wireframe" view of the illustration (Fig. 4.5, center).

- Select the line with the Node tool

- Use the shortcut key X to toggle between the *Fill* color and the *Line* color. Change the *fill* to *empty* (by left-clicking the color ⊠), and the *color* to the color you want.

- Adjust the line-width and the line-style

And for handling all the text in an own layer:

- In the Layers panel, select Add Layer

- Call the original layer "Figure", and the new layer "Labels"

- Select all the Text elements with the Move tool, group them with CTRL+G, and label the group in the Layers-panel Text.

- Move/drag this group from "Figure" to "Labels"

- Lock "Figure"

- Select "Labels", and make all the desired text adjustments.

**Inkscape**

In Inkscape the procedure is similar to the other programs. Inkscape is free and open source and gets the job done. But it is in my opinion less intuitive and less convenient to work with than commercial programs.

- `File | Import` → select `sine.svg`

- Select the figure with the `Select and Transport`-tool, right-click and `Ungroup` it

- Select the chosen line, and type `Ctrl+Shift+F` to adjust the `Fill and Stroke`-properties

- Type `Ctrl+Shift+L` to get to the layers, and add a new Layer for the Text, and place it above `Layer 1`

- To move the existing labels from `Layer 1` to the `Text`-layer, `Group` all the labels together, select them, and type `Shift+PageUp`.

## 4.5 Displaying Data Sets

Below a short list of the most important *matplotlib* graphics types is presented.

### 4.5.1 Plots of Data with One Variable

The following examples all have the same format. Only the "Plot-command" line changes.

```
# Import standard packages
import numpy as np
import matplotlib.pyplot as plt
import pandas as pd

# Generate the data
x = np.random.randn(500)

# Plot-command start --------------------
plt.plot(x, '.')
# Plot-command end ----------------------

# Show plot
plt.show()
```

**Scatter Plots**

This is the simplest way to represent "univariate" data, i.e. data with a single variable: just plot each individual data point. The corresponding plot-command is

```
plt.plot(x, '.')
```

Things to consider when generating scatter plots:

- If the data are not part of a sequence (e.g. data as a function of time), do *not* connect data points with a line.

- If there are few data points, it might be aesthetically more pleasing to use 'o' or '*' instead of '.' as plot-symbol, leading to larger dot sizes (Fig. 4.7).

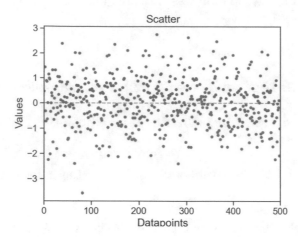

Figure 4.7: Scatter plot (from the code-quantlet ("CQ") show_plots.py).

Note: In cases where we only have few discrete values on the x-axis (e.g. *Group1, Group2, Group3*), it may be helpful to spread overlapping data points slightly (also referred to as *"adding jitter"*) to show each data point. An example can be found at http://stanford.edu/~mwaskom/ software/seaborn/generated/seaborn.stripplot.html.

**Histograms**

Histograms provide a good first overview of the distribution of data. The box-width is arbitrary, and the smoothness of the histogram depends on the chosen box-width. The histogram can be represented as a *frequency histogram*, by simply counting the number of samples in each box. By using the option `density=True` of the command `plt.histogram` histograms can be "normalized", which corresponds to dividing the frequency counts by the total number of samples. In that case the value of each box corresponds to the probability of finding a data value in the corresponding data range, and the sum over all values is exactly 1 (Fig. 4.8).

```
plt.hist(x, bins=25, density=True)
```

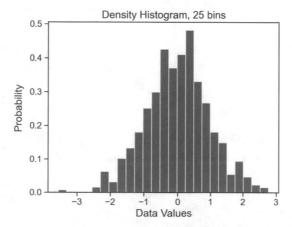

Figure 4.8: A *density histogram* indicates the probability to find a single data sample in the corresponding bin with a bar (from CQ show_plots.py).

### Error Bars

Error-bars are a common way to show mean value and variability when comparing measurement values. Note that it always has to be stated explicitly if the error-bars correspond to the *standard deviation* or to the *standard error* of the data. *Standard errors* (see Sect. 7.2.2) make it easy to discern statistical differences between groups: when error bars for the standard errors for two groups overlap, one can be sure the difference between the two means is not statistically significant ($p > 0.05$). (However, the opposite is not always true!)

The following commands also show how to replace the tick labels on the x-axis with strings (Fig. 4.9):

```
weight = {'USA':89, 'Austria':74}
weight_SD_male = 12
plt.errorbar([1,2], weight.values(), yerr=weight_SD_male * np.r_[1,1],
  capsize=5, LineStyle='', marker='o')
plt.xlim([0.5, 2.5])
plt.xticks([1,2], weight.keys())
plt.ylabel('Weight [kg]')
plt.title('Adult male, mean +/- SD')
```

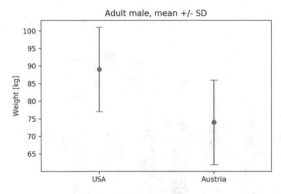

Figure 4.9: Displaying mean and variability for different groups (from CQ show_plots.py).

### Box Plots

Box plots are frequently used in scientific publications to indicate values in two or more groups. The bottom and top of the box indicate the "first quartile" (i.e. the value larger than 25% of the data) and "third quartile" (i.e. the value larger than 75% of the data), respectively. The line inside the box shows the median (the value larger than 50% of the data). In other words, the box contains 50% of the data samples (Fig. 4.10). (More information on box plots can be found in the Chap. 7.)

```
plt.boxplot(x, sym='*')
```

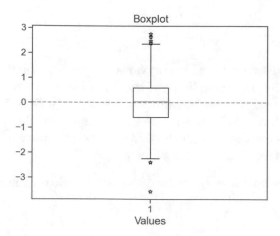

Figure 4.10: Box plot (from CQ `show_plots.py`).

## Grouped Bar Charts

For some applications the plotting abilities of *pandas* can facilitate the generation of useful graphs, e.g. for grouped bar plots (Fig. 4.11):

```
df = pd.DataFrame(np.random.rand(7, 3),
        columns=['one', 'two', 'three'])
df.plot(kind='bar', grid=False)
```

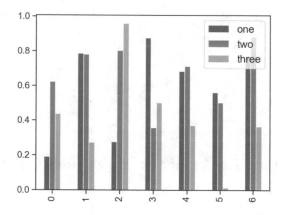

Figure 4.11: Grouped barplot, produced with *pandas* (from CQ `show_plots.py`).

## Pie Charts

*Pie charts* can be generated with a number of different options. Here I use colormaps from the statistics visualization package *seaborn*, which is commonly imported as `sns`. Among many statistical visualization features, *seaborn* also offers a number of practical color maps (https://seaborn.pydata.org/tutorial/color_palettes.html) (Fig. 4.12).

```
import seaborn as sns
import matplotlib.pyplot as plt

txtLabels = 'Cats', 'Dogs', 'Frogs', 'Others'
fractions = [45, 30, 15, 10]
offsets =(0, 0.05, 0, 0)
```

```
plt.pie(fractions, explode=offsets, labels=txtLabels,
        autopct='%1.1f%%', shadow=True, startangle=90,
        colors=sns.color_palette('muted') )
plt.axis('equal')
```

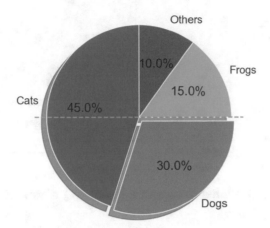

Figure 4.12: "Sometimes it is raining cats and dogs" (from CQ show_plots.py).

**Programs: Data Display**

🐍 python **Code[3]:** show_plots.py contains the Python code to generate the plots in this section.

### 4.5.2 Plots of Data with Two or More Variables

There is always a trade-off between simplicity and information density. Care should be taken to keep plots that put more information into a single graph understandable to the reader.

**Scatter Plots**

Scatter plots of two-dimensional data ("bivariate data") can add additional information through the symbol size. This type of plot is still very straight-forward to understand.

In order to make the plot reproducible, I specify a "seed" for the generation of the random numbers with the command np.random.seed (Fig. 4.13):

```
np.random.seed(12)
data = np.random.rand(50, 3)
plt.scatter(data[:,0], data[:,1], s=data[:,2]*300)
```

---

[3]https://github.com/thomas-haslwanter/sapy/blob/master/src/code_quantlets/show_plots.py.

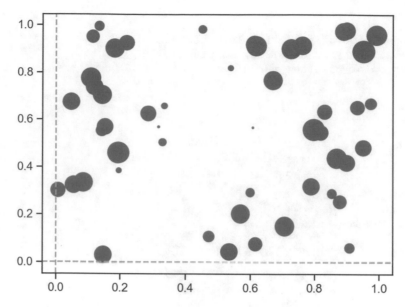

Figure 4.13: Scatter plot with scaled datapoints (from CQ `show_plots.py`).

## 3D Plots

Data of three or more dimensions are sometimes called "multivariate data". For 3D plots *matplotlib* requires that separate modules have to be imported, and that axes for 3D plots are declared explicitly.

**Note:** Use 3D plots sparingly, thought, since it is usually almost impossible to estimate quantitative relationships from 3D plots.

Once the 3D plot axis is correctly defined the plotting is straightforward. Here two examples (Fig. 4.14):

```
# imports specific to the plots in this example
import numpy as np
from matplotlib import cm
from mpl_toolkits.mplot3d.axes3d import get_test_data

# Twice as wide as it is tall.
fig = plt.figure(figsize=plt.figaspect(0.5))

#---- First subplot
# Note that the declaration "projection='3d'"
# is required for 3d plots!
ax = fig.add_subplot(1, 2, 1, projection='3d')

# Generate the grid
X = np.arange(-5, 5, 0.1)
Y = np.arange(-5, 5, 0.1)
X, Y = np.meshgrid(X, Y)

# Generate the surface data
R = np.sqrt(X**2 + Y**2)
Z = np.sin(R)

# Plot the surface
surf = ax.plot_surface(X, Y, Z, rstride=1, cstride=1,
        cmap=cm.GnBu, linewidth=0, antialiased=False)
```

```
ax.set_zlim3d(-1.01, 1.01)

fig.colorbar(surf, shrink=0.5, aspect=10)

#---- Second subplot
ax = fig.add_subplot(1, 2, 2, projection='3d')
X, Y, Z = get_test_data(0.05)
ax.plot_wireframe(X, Y, Z, rstride=10, cstride=10)

outfile = '3dGraph.jpg'
plt.savefig(outfile, dpi=200)
print(f'Image saved to {outfile}')
plt.show()
```

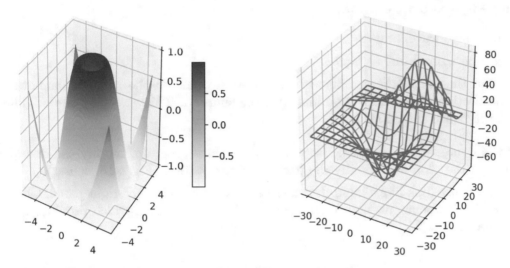

Figure 4.14: Two types of 3D graphs. **Left:** surface plot. **Right:** wireframe plot (from CQ show_plots.py).

# 4.6   Exercises

1. **Plotting Data** From the command-line, create two cycles of a noisy sine wave with the following properties: amplitude = 1, frequency = 0.3 Hz, sample_rate = 100 Hz. Add Gaussian random noise with a standard deviation of 0.5 to these data.

   - Plot the data, label the x- and the y-axis, and add a title to the plot.

   - When this works, take your command history, clean it up, and create a Python function that

     - takes the number of cycles and the frequency as input
     - sets the remaining parameters as above
     - shows the resulting graph

   - Set a breakpoint in the function somewhere after the amplitude has been defined, and inspect the workspace variables at that point. Modify the amplitude to *2* and continue the function.

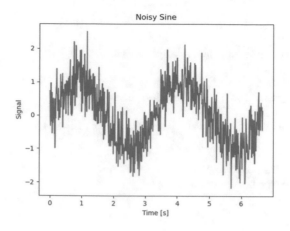

2. **Modifying Figures** Start out with

   - plotting a sine wave, and

   - adding a horizontal line at *y=0.8*.

   - Then find the *matplotlib* command to annotate a point on the plot, and annotate the intersection of the horizontal line with the sinusoid. This gives you the original *matplotlib* figure (Fig. 4.15, left).

   - Save the figure in SVG-format. Use a vector graphics program, modify the line attributes of the plot, and modify the text of the annotation (Fig. 4.15, center)

   - Make the figure look like a hand-drawn sketch, as indicated in the right panel. Check the *matplotlib* documentation for the command `plt.xkcd()`, which provides that functionality.

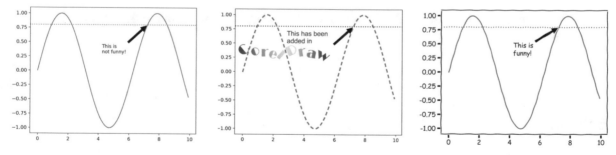

Figure 4.15: **Left:** original *matplotlib* figure, with annotation. **Center:** SVG figure, modified in a vector graphic program. **Right:** same as left figure, but with `plt.xkcd()`.

# Chapter 5

# Data Filtering

Often 1-dimensional data have been sampled at equal increments. Take for example position data, recorded at a frequency of 1 kHz. Most of this chapter focuses on the analysis of such regularly sampled data. The first section establishes the basic terminology. The second section introduces "linear time invariant" filters as well as non-linear filters. The third section discusses impulse-, step-, and frequency-responses of these filters, and the problem of time delay. The fourth chapter presents different solutions to the most frequently occurring filtering tasks: smoothing, differentiation, and integration of signals. An advanced outlook presents common but more complex ways of smoothing data, which can also be applied to irregularly sampled data. We will briefly discuss *local regression smoothing* ("LOESS-filtering"), *B-splines*, and *kernel density estimations* ("KDE-plots"). The last section shows how the same concepts that underlie FIR-filters and morphological filters can also be applied to higher dimensional data such as images.

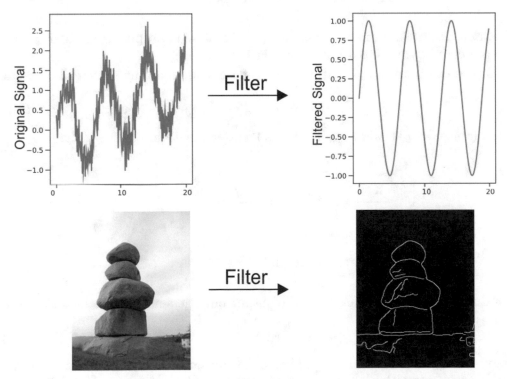

Figure 5.1: Filters applied to 1D (top) and 2D (bottom) signals.

The first steps in data analysis often involves one or more of the following tasks:

- smoothing the data

- differentiation

- low-, high-, or band-pass filtering

- elimination of outliers (morphological filters)

- integration.

For example, Fig. 5.1 (top) shows a 1-dimensional (1D) data signal. The original data contain not only the signal, a sine-wave, but also a low-frequency drift and high-frequency noise. Ideally, a filter should eliminate all the undesired components. In other cases, we may want to use a filter to extract the outlines from 2-dimensional (2D) signals, for example images (Fig. 5.1, bottom).

## 5.1   Transfer Functions

To analyze signals you have to modify them. If the incoming data are called $x$ and the resulting/outgoing data $y$, this can be indicated schematically with the diagram in Fig. 5.2.

Figure 5.2: Transfer function.

A *transfer function* describes the mapping of an input signal to an output signal. In diagrams it is commonly represented by a labeled box, as in Fig. 5.2.

For the trivial case of a pure amplification, a triangle is commonly used instead of a box (Fig. 5.3):

$$y = g * x \qquad (5.1)$$

The change in amplitude $g$ is called "gain" of the transfer function.

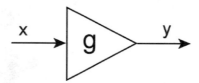

Figure 5.3: Amplification.

For discrete valued signals this means that the output only depends on the instantaneous input:

$$y(n) = g * x(n) \qquad (5.2)$$

When the gain of a system covers many orders of magnitude, the system behavior is visualized more easily on a logarithmic scale. In that case the relative change in amplitude is commonly referred to in *deciBel (dB)*:

$$attenuation = 20 * log_{10} \left( \frac{a_{out}}{a_{in}} \right) \text{dB.} \qquad (5.3)$$

For example, an amplitude reduction by a factor of *2* corresponds to an attenuation of 6 dB, and a reduction by a factor of 10 to an attenuation of 20 dB.

## 5.2 Filter Types

In this section the most common filter types are introduced: *Finite Impulse Response (FIR)* filters and *Infinite Impulse Response (IIR)* filters, both of which are *Linear Time Invariant (LTI* filters*)*, and *morphological filters* such as the median filter.

### 5.2.1 Linear Time Invariant (LTI) Filters

In most cases linear filters are *linear time invariant (LTI)* filters. A necessary condition for the linearity is that "doubling the input doubles the output". Or more formally:

**Linearity** The definitive test for a linear system is that if input $x_1(t)$ produces output $y_1(t)$ and $x_2(t)$ produces $y_2(t)$, then the input $a * x_1(t) + b * x_2(t)$ must produce the output $a * y_1(t) + b * y_2(t)$. These are the *superposition* and *scaling* properties of linear systems. With

$$x_k(t_i) \rightarrow y_k(t_i), \ k = 1, 2 \tag{5.4}$$

we get

$$a_1 * x_1(t_i) + a_2 * x_2(t_i) \rightarrow a_1 * y_1(t_i) + a_2 * y_2(t_i) \ . \tag{5.5}$$

**Time invariance** This means that the system coefficients do not change for the period of the investigation.

An example of a linear system is $y(t_i) = 3 * x(t_i) + 2 * x(t_{i-1})$. An obvious example for the opposite, a non-linear system, would be $y(t_i) = x(t_i)^2$. A less obvious example of a system that would *not* fulfill our definition of linearity is $y(t_i) = x(t_i) + c$ : while this system describes a line in space, doubling the input does not double the output, with the exception of the trivial case $c = 0$.

### 5.2.2 Finite Impulse Response (FIR) Filters

The superposition and scaling properties of linear filters allow two independent ways of analyzing LTI filters: the *input view* starts out with a given input, and investigates the effects of that input on the subsequent outputs; in contrast, the *output view* analyzes which parts of the input have contributed to a given output value.

We start with the output view of FIR filters, represented by Eq. (5.6). (The corresponding input view is the *impulse response*, which evaluates the effect of a change of a single input sample on the output. The impulse response is described in Sect. 5.3 below.)

For an FIR filter of order $k$, the output is determined by the weighted sum of the last $k + 1$ inputs, with the weights $w(k)$:

$$y(n) = \sum_{i=0}^{k} w_i * x(n - i) = \tag{5.6}$$
$$= w_0 * x(n) + w_1 * x(n - 1) + \cdots + w_k * x(n - k)$$

This can be seen as a *moving window filter*, which is moved over the incoming data points from beginning to end (Fig. 5.4), and is called "finite impulse response (FIR)" filter. The coefficients in Eq. (5.6) start with *0*, reflecting conventions in the area of *Digital Signal Processing (DSP)*.

**Example: Moving Average**

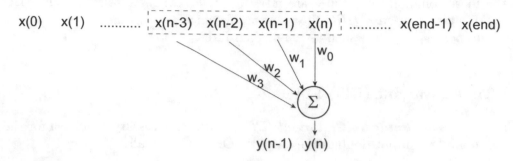

Figure 5.4: FIR filter, for online analysis.

If the filter values are chosen as

$$w_i = \frac{1}{k+1}, \quad i = 0 : k \tag{5.7}$$

the filter averages over the last k+1 data points. For example, $k = 2$ indicates a filter that averages over the last three data points. This is implemented with the weight vector

$$\mathbf{w} = \begin{bmatrix} \frac{1}{3} & \frac{1}{3} & \frac{1}{3} \end{bmatrix} \tag{5.8}$$

For example, for the 11th data point ($n = 10$, remember that the counting starts at $0$) this gives the output value

$$y(10) = w_0 * x(10) + w_1 * x(9) + w_2 * x(8) = \frac{x(10) + x(9) + x(8)}{3} \tag{5.9}$$

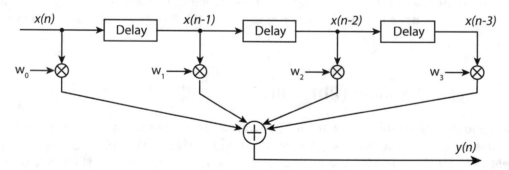

Figure 5.5: DSP representation of an FIR filter. A delay by one sample is often indicated with $z^{-1}$, because this corresponds to the *z-transform* of a one-sample-shift of the input (see Sect. 10.5.2).

We can implement Eqs. (5.6) and (5.8), smoothing with an averaging filter, in Python with the command `signal.lfilter` (Fig. 5.5):

```
import numpy as np
from scipy import signal
w = np.ones(3)/3
y = signal.lfilter(w, 1, x)
```

where **w** is a vector containing the weights, and **x** a vector containing the input. The parameter *"1"* for the command `signal.lfilter` will be explained when we discuss the more general IIR-filters below.

Equation (5.6) is sometimes also called a *convolution* of the signal $\mathbf{x}$ with a *convolution-kernel* $\mathbf{w}$, and is then written as

$$y(n) = w(k) * x(n) \tag{5.10}$$

In other words:

---

Application of an FIR-filter with weights $\mathbf{w}$ is equivalent to a convolution of the signal with $\mathbf{w}$.

---

### 5.2.3   Infinite Impulse Response (IIR) Filters

While the output of a FIR filter only depends on the incoming signal (Eq. (5.6)), the general output of a filter may also depend on the $m$ most recent values of the output signal:

$$y(n) + a_1 * y(n-1) + \cdots + a_m * y(n-m) = b_0 * x(n) + b_1 * x(n-1) + \cdots + b_k * x(n-k) \tag{5.11}$$

or equivalently

$$\sum_{j=0}^{m} a_j * y(n-j) = \sum_{i=0}^{k} b_i * x(n-i) \tag{5.12}$$

where $a_0 = 1$. The coefficients $a_i$ and $b_j$ uniquely determine this type of filter, which is commonly referred to as "infinite impulse response" (IIR) filter (Figs. 5.6) and 5.7. The "order" of the filter is defined as the larger of $m$ and $k$ in Eq. (5.12). The feedback character of these filters can be made more obvious by reorganizing the equation:

$$y(n) = [b_0 x(n) + b_1 x(n-1) + \cdots + b_k x(n-k)] - [a_1 y(n-1) + \cdots + a_m y(n-m)] \tag{5.13}$$

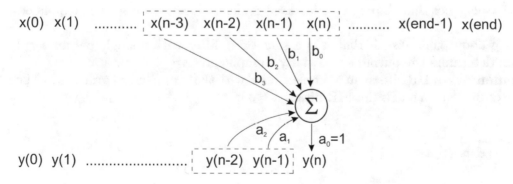

Figure 5.6: Infinite impulse response filter (IIR filter).

While it is more difficult to find the desired coefficients $(\boldsymbol{a}, \boldsymbol{b})$ for IIR filters than for FIR filters, IIR filters have the advantage that they are computationally more efficient, and achieve a better frequency selection with fewer coefficients.

**Example: Exponential Averaging Filter**

**Exponential Decay**   When the change in a signal is proportional to the magnitude of the signal, the signal changes exponentially:

$$y_n = \alpha * y_{n-1} \tag{5.14}$$

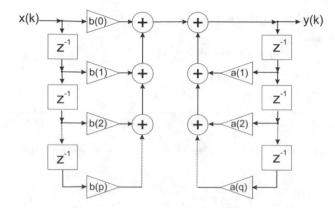

Figure 5.7: Typical DSP-representation of an IIR Filter. This implementation is called *Direct Form I* representation of the filter. (For other presentations, see Smith 2007). The time delay is labeled "$z^{-1}$", corresponding to the z-transform of a time delay by the sample period.

For $\alpha > 1$ the signal grows exponentially, and for $\alpha < 1$ it decreases exponentially. For example, for $\alpha = 1/2$ we get (Fig. 5.8)

$$y_i = y_0 * \left[ 1, \frac{1}{2}, \frac{1}{4}, \frac{1}{8}, \cdots \right]$$

**Combination with Input Signal** When the new output state $y(n)$ is a weighted combination of the input signal $x(n)$ and feedback from the previous output state $y(n-1)$, the resulting filter can be written as

$$y(n) = \alpha * x(n) + (1 - \alpha) * y(n-1) \tag{5.15}$$

This filter is called either "exponential averaging filter" or "leaky integrator". From Eq. (5.12) one can see that bringing all the terms with $y$ to the left gives us the feedback coefficients $\mathbf{a} = [1, -(1 - \alpha)]$, and bringing all terms containing an $x$ to the right the feed forward coefficients $\mathbf{b} = [\alpha]$. One also sees that the exponential decay is the impulse response of an exponential averaging filter.

The nice thing about this filter is that it is a smoothing filter with a single parameter: by tuning $\alpha$ one can determine the output's sensitivity to input noise.

**Python Implementation** IIR filters can be implemented with the command `signal.lfilter` by specifying the coefficients $\mathbf{a}$ and $\mathbf{b}$ (for FIR-filters: $a = 1$):

```python
import numpy as np
from scipy import signal

# Dummy input_data and parameter
x = np.random.randn(1000)
alpha = 0.5

# IIR filter coefficients for exponential averaging filter
a = [1, -(1-alpha)]
b = [alpha]

# Apply the filter
y = signal.lfilter(b,a,x)
```

### Difference Between FIR and IIR Filters

The advantage of FIR filters is their simplicity; the advantage of IIR filters is the sharper frequency responses that can be achieved with a given filter order. The two names *finite impulse response filter* and *infinite impulse response filter* are derived from the differing behavior of each type of filter to an impulse input. To demonstrate this graphically, we can implement an example of each filter type:

Listing 5.1: fir_vs_iir.py

```python
""" Show the effect of an FIR- and an IIR-filter on an impulse """

# Import the standard packages
import numpy as np
import matplotlib.pyplot as plt
from scipy import signal

# Generate the impulse and the time-axis
xx = np.zeros(20)
xx[5] = 1
tt = np.arange(20)

# Put the results into a Python-dictionary
data = {}
data['before'] = xx
data['after_fir'] = signal.lfilter(np.ones(5)/5, 1, xx)
data['after_iir'] = signal.lfilter([1], [1, -0.5], xx)

# Show the results
plt.plot(tt, data['before'], 'o', label='input', lw=2)
plt.plot(tt, data['after_fir'], 'x-', label='FIR-filtered', lw=2)
plt.plot(tt, data['after_iir'], '.:', label='IIR-filtered', lw=2)

# Format the plot
plt.xlabel('Timesteps')
plt.ylabel('Signal')
plt.legend()
plt.xticks(np.arange(0, 20, 2))
plt.gca().margins(x=0, y=0.02)

# Save and show the image
out_file = 'FIRvsIIR.jpg'
plt.savefig(out_file, dpi=200, quality=90)
print(f'Image saved to {out_file}')

plt.show()
```

This code produces Fig. 5.8. In this graph we can see:

- the time delay of the FIR filter

- the finite effect of an impulse on FIR-filtered data

- the instant response of the IIR filter

- the infinite effect of an impulse on IIR-filtered data

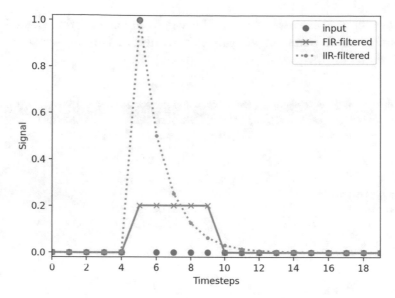

Figure 5.8: Comparison of FIR and IIR filter behavior. The solid line shows the impulse response of an FIR filter, a *5-point averaging filter*, and has only 5 values different from 0. In contrast, the output of an *exponential averaging filter* (which is an IIR-filter) never returns to zero (from Listing 5.1.).

### 5.2.4   Morphological Filters

FIR-filters and IIR-filters are both *linear filters*, since the filter coefficients enter the transfer function only linearly. While such linear filters are good at eliminating noise that has a Gaussian distribution, it fails for other tasks, e.g. for removing extreme outliers. Such spikes in the data can be caused e.g. by faulty sensors or by loose connections in the experimental setup. For such tasks so-called "morphological filters" can be useful. Instead of using weighted averages of inputs/outputs, those morphological filters use data features such as the

- minimum

- maximum

- median

- range

of the elements within a data window. For example, for removing extreme outliers a *median filter* offers a better noise suppression than linear filters, as the following example demonstrates. In Fig. 5.9 the signal has two outliers, one at *t = 5*, and one at *t = 15*. The averaging filter has been adjusted to compensate for the delay, and both averaging and median filter have a window size of 3.

Listing 5.2: median_filter.py

```
""" Demonstration of linear and non-linear filters on data with extreme outliers
    """

# Import the standard packages
import numpy as np
import matplotlib.pyplot as plt
from scipy import signal
```

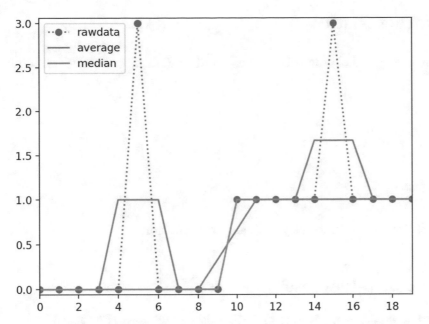

Figure 5.9: Data with extreme outliers (blue, dotted), average filtered (orange, solid), and median filtered (green, solid) (from Listing 5.2).

```python
# Create the data
x = np.zeros(20)
x[10:] = 1

# Add some noise-spikes
x[[5,15]] = 3

# Median filter the signal
x_med = signal.medfilt(x, 3)

# Average filtered data
b = np.ones(3)/3
x_filt = signal.lfilter(b, 1, x)

# Plot the data
plt.plot(x, 'o', linestyle='dotted', label='rawdata')
plt.plot(x_filt[1:], label='average')
plt.plot(x_med, label='median')

plt.xlim([0, 19])
plt.xticks(np.arange(0,20,2))
plt.legend()

ax = plt.gca()
ax.margins(x=0, y=0.02)

# Save and show the image
out_file = 'MedianFilter.jpg'
plt.savefig(out_file, dpi=200, quality=90)
print(f'Image saved to {out_file}')

plt.show()
```

## 5.3    Filter Characteristics

There are three common ways to characterize the effect of an LTI filter on a given input signal:

- Impulse response

- Step response

- Frequency response

All three representations contain exactly the same information.

### 5.3.1    Impulse- and Step-Response

An *impulse* is an input where one value is 1, and the rest are all zero (Fig. 5.10).

$$x(i) = 1 \text{ for } i = k$$
$$x(i) = 0 \text{ for } i \neq k$$

In contrast, the *step* jumps from zero to one and remains there.

> The outputs of the impulse response (around point $k$) are exactly the weight coefficients **w** of the FIR-filter.

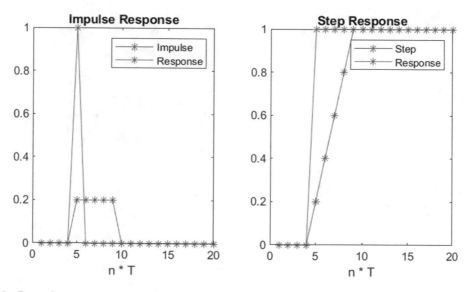

Figure 5.10: Impulse response and step response, for a 5-point moving average filter. (From F5_filter_characteristics.py, Appendix A.)

### 5.3.2    Frequency Response

An important property of LTI transfer functions is that a sine-input with a given frequency always leads to a sine output with the same frequency, with only phase and/or amplitude modified (see Fig. 5.11).

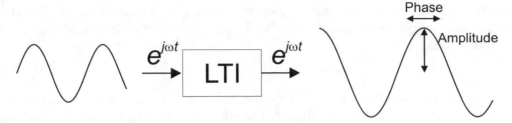

Figure 5.11: A sine input to an LTI transfer function always leads to a sine output with the same frequency. However, the *amplitude* and the *phase* can change.

Amplitude and phase of the transfer gain can conveniently be expressed as a single complex number: the amplitude corresponds to the length of the complex number, and the phase to the angle (Fig. 5.12).

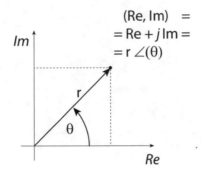

Figure 5.12: Representation of the complex number $(Re + j*Im)$ in polar coordinates $(r, \theta)$. $r$ can characterize the gain of a transfer function, and $\theta$ the corresponding phase shift.

This can be used to characterize the filter response. As an example, Fig. 5.13 shows the normalized frequency response of a 5-point averaging filter, produced with the command `scipy.signal.freqz`. Note that the gain is commonly represented on a logarithmic scale in $dB$, as defined in Eq. (5.3).

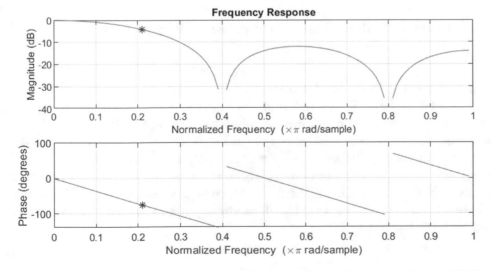

Figure 5.13: Frequency response for a 5-point moving average filter. The gain and phase value of the frequency used in Fig. 5.14 are indicated with a "*". (From `F5_filter_characteristics.py`, Appendix A.)

The x-axis of Fig. 5.13 contains the *normalized frequency*, where 1 corresponds to the *Nyquist frequency*. According to the *Nyquist–Shannon sampling theorem*, information in a signal can only be faithfully reproduced up to

$$f_{Nyquist} = \frac{rate}{2} \tag{5.16}$$

Higher signal frequencies introduce artifacts, as show in Fig. 1.4. For example, for a sample rate of 1 kHz, the Nyquist frequency would be 500 Hz. The effect of the 5-point averaging filter on a sine-input with 105 Hz, for a sample rate of 1 kHz, is indicated in Fig. 5.14. The corresponding gain and phase values in Fig. 5.13 are marked with ∗.

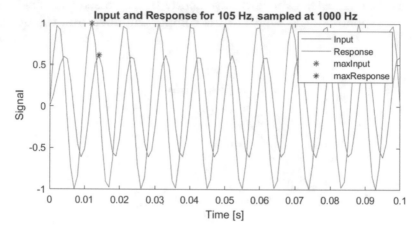

Figure 5.14: Effect of a 5-point averaging filter on a frequency of 105 Hz, sampled at 1 kHz. The indicated second maxima of input and output are used in the attached program `F5_filter_characteristics.py` to numerically estimate gain and phase (from `F5_filter_characteristics`, Appendix A).

For continuous systems, the frequency response can also be obtained with the function `scipy.signal.bode` (see Sect. 10.4).

### 5.3.3   Artifacts in Causal Filters, and Non-causal Filters

#### Causal Filters

All filters discussed so far show some "initial transients", i.e. artifacts that are caused by the initialization values of FIR and IIR filters and that only appear at the start of the signal. In addition, they always have a time delay of the output relative to the input (Fig. 5.15). The initial transients are due to the fact that for $i < length_{Filter}$ (in Fig. 5.15 for the first four points), the filter window is longer than the input data already available. By convention, the corresponding missing input values are set to 0, and it is best to ignore the transients in data analysis. The second consistent effect is a delay of the output data, which is especially relevant for real-time data analysis: real-time filters can only act on the data already available, leading to a constant delay of the output relative to the input.

#### Non-causal Filters

**Centered Analysis Window**  For offline analysis it is often more convenient to use a window that is centered about the current position (Fig. 5.16). This eliminates the problem of time delays in the filtered output signal $y$ relative to the input signal $x$. In the language of DSP this is called a *non-causal filter*, since future points are included in the evaluation of $y_i$—which is not possible in the real-time analysis of data.

Transient
Delay
Underestimated extrema
cos
filtered

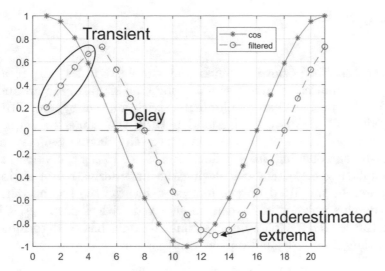

Figure 5.15: Effects of an FIR-filter, here a 5-point averaging filter. While the initial transient and the time-delay are general features of causal filters, the underestimation of extrema is specific to the averaging filter, and can be avoided for example by using a Savitzky–Golay filter (see next section).

$$y_n = \sum_{m=-k}^{k} w_m x_{n+m} \tag{5.17}$$

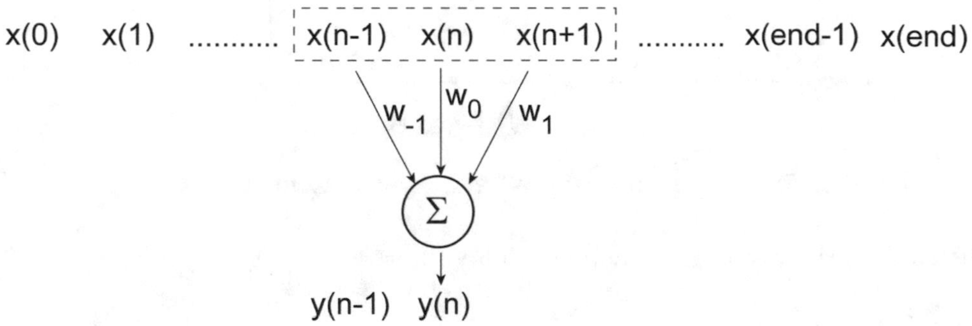

x(0) x(1) .......... x(n-1) x(n) x(n+1) .......... x(end-1) x(end)
w-1 w0 w1
Σ
y(n-1) y(n)

Figure 5.16: Filter using a centered window (offline analysis).

**Filtfilt** A different non-causal filter, `scipy.signal.filtfilt`, avoids this delay by running the filter over the data twice: once front-to-back, and once back-to-front. Be careful, however, as this double application of the filter also changes the filter characteristics!

## 5.4 Applications

This section discusses *smoothing*, *differentiation*, and *integration* of signals. Since the so-called "Savitzky–Golay filter", an FIR filter, can be used for both smoothing and differentiation, we begin with the presentation of that filter.

Bandpass-filters, i.e. filters that only transmit data in a limited frequency range, are not specifically discussed here. However, it should be noted that the Butterworth filter presented below (Sect. 5.4.2) can be used not only for smoothing, but also for high-pass and band-pass filtering.

### 5.4.1   Savitzky–Golay Filter

**Definition**

The moving average filter described above is quick and simple, but its output is not very accurate. For example, it systematically underestimates the true signal value around the peak of a signal (see Fig. 5.15). This problem can be eliminated with another type of FIR-filter, the *Savitzky–Golay filter* (Savitzky and Golay 1964; Madden 1978; the numerical implementation is well described in Press et al. 2007). To smooth a curve, the Savitzky–Golay filter fits a polynomial of order $q$ to the surrounding $2m + 1$ data points, and uses the value at its center for the output at this point (Fig. 5.16). For calculating the first derivative of an input signal the inclination at the center of the fitted polynomial is taken, for the second derivative the curvature etc. Surprisingly, this polynomial-fit can be achieved with a simple FIR filter. One of the best things about the Savitzky–Golay filter is that one only has to determine the filter coefficients **w** once, since they do not depend on the input signal.

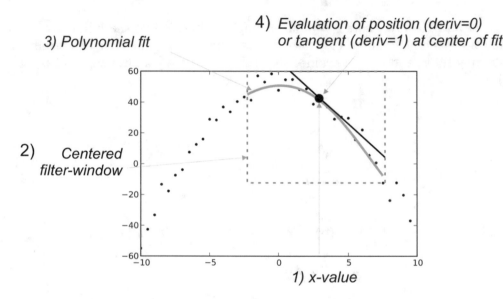

Figure 5.17: The principle of the Savitzky–Golay filter.

Figure 5.17 shows the principle of the Savitzky–Golay filter:

1. For a given x-value . . .

2. a symmetric window about x is selected, then . . .

3. the best polynomial fit to these data is calculated, and . . .

4. for smoothing, the center-point of the fitted curve is taken;
   for the 1st derivative, the tangent at that location;
   for the second derivative, the curvature, etc.

As a result, the Savitzky–Golay filter requires the following parameters:

- *order of the polynomial fit* (e.g., "2" for quadratic fits)

- *size of the data window* (must be odd, e.g., "9")

- *order of derivative* (e.g., "0" for smoothing, "1" for 1st derivative)

- *sampling rate* (in Hz, e.g., "100"; not required for smoothing)

Two commands from `scipy` implement all the required functionality:

For smoothing as well as for derivatives the command `savgol_filter` from the package `scipy.signal` can be used. And should the filter coefficients **w** be required explicitly, they can be obtained for general polynomials and window sizes with the command `savgol_coeffs`.

### Examples

**Example 1** The smoothing of data with the best-fit second-order polynomial, fit to the data in a five point window, can be achieved with an FIR filter with the following coefficients:

```
from scipy import signal

w = signal.savgol_coeffs(window_length=5, polyorder=2, deriv=0)
#w = [-0.09, 0.34, 0.49, 0.34, -0.09]
filtered = signal.lfilter(w,1,in_data)
```

or equivalently

```
filtered  = signal.savgol_filter(in_data, window_length=5, polyorder=2, deriv=0)
```

**Example 2** The "cubic differentiator" presented below (Eq. 5.21) is equivalent to

```
signal.savgol_filter(in_data, window_length=5, polyorder=3, deriv=1)
```

### Comments

**Advantages of Savitzky–Golay filters** They are efficient; they are very convenient to calculate higher derivatives; and smoothing and derivations can be done simultaneously.

**Disadvantages of Savitzky–Golay filters** They don't have a crisp frequency response. In other words, the gain decreases only gradually as frequency increases. For example, if only frequency components below $200\,\mathrm{Hz}$ should pass through the filter, other filtering techniques are preferable, for example Butterworth filters (see Sect. 5.4.2). Also, if the ideal signal characteristics of the true signal are known, a *Wiener filter*, though more complex to apply, may produce better results (Wiener 1942).

## 5.4.2   Smoothing of Regularly Sampled Data

A number of smoothing filters have already been presented:

- Moving average filter (see Sect. 5.2.2)

- Exponential averaging filter (see Sect. 5.2.3)

- Median filter (see Sect. 5.2.4)

- Savitzky–Golay filter (see Sect. 5.4.1)

Another important smoothing filter is the *Butterworth low-pass filter*:

**Butterworth Low-Pass Filter**

The `scipy.signal` command `butter` provides the $[b, a]$ coefficients of an IIR filter corresponding to *Butterworth filters*. The Butterworth filter is a type of signal processing filter designed to have as flat a frequency response as possible in the pass band. It is therefore also referred to as a "maximally flat magnitude" filter. It can be used as a low-pass, a high-pass, or a band-pass filter.

For example, for a sampling rate of 1 kHz (and thus a Nyquist frequency of 500 Hz), a Butterworth low-pass filter with a 3 dB cut-off frequency of 40 Hz and a filter-order of 5 can be obtained with:

```
from scipy.signal import butter, lfilter

# Dummy input_data
x = np.random.randn(1000)

nyq = 500
cutoff = 40
filter_order = 5
b,a = butter(filter_order, cutoff/nyq)

filtered = lfilter(b, a, x)
```

**Warning:** Be careful with low filter frequencies and higher order ($n \geq 4$) Butterworth filters where the $[b, a]$ syntax may lead to numerical problems due to round-off errors. In that case the the "SOS" (second-order sections) syntax, which is the recommended form for IIR-filters, or the "ZPK" (zero-pole-gain) syntax should be used. Descriptions of these representations can be found e.g. in (Smith 2007).

**Notes**

- Depending on the application, other frequency responses might be preferable. For example, *Chebyshef filters* provide sharper frequency responses than Butterworth filters, and *Bessel filters* have the advantage that they show no overshoot in the time domain.

- For fitting data to a *parametric model* it is almost always better to use raw data than pre-smoothed data, since smoothing already discards available information.

### 5.4.3   Differentiation

Differentiation of data is a commonly occurring task. It can be used to find velocity and acceleration corresponding to a given position signal, to find extreme values, to determine tangents to curves, and for many other applications. In this section we present three possible ways to determine a derivative numerically.

**First-Difference Differentiation**

A differentiation of an incoming signal is given by (Fig. 5.18, top)

$$y(n) = \frac{\Delta x}{\Delta t} = \frac{x(n) - x(n-1)}{\Delta t} \tag{5.18}$$

This gives the filter weights for an FIR-filter

$$\mathbf{w} = [1, \ -1] / \Delta t \tag{5.19}$$

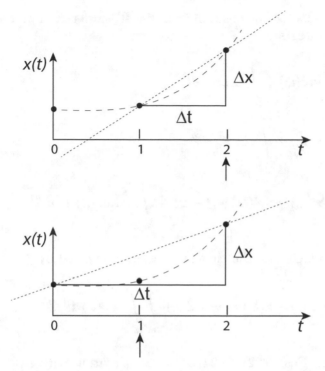

Figure 5.18: **Top:** first-difference differentiator. **Bottom:** central-difference differentiator.

## Central-Difference Differentiator

For offline analysis a centered filter may be preferable (Fig. 5.18, bottom):

$$\mathbf{w} = [1, \; 0, \; -1] * \frac{1}{2 * \Delta t} \tag{5.20}$$

## Cubic Differentiator

We can also differentiate a curve by taking two samples before and after each point, and taking the slope of the best cubic fit to the central data point. (Note that this is a special case of the Savitzky-Golay filter.) This can be achieved with a weight vector of:

$$\mathbf{w} = [1, \; -8, \; 0, \; 8, \; -1] * \frac{1}{12 * \Delta t} \tag{5.21}$$

## Specialized Differentiators

To optimize noise and frequency responses, additional differentiation algorithms have been developed which won't be discussed here:

- Lanczos differentiator (Lanczos 1961)

- Parks–McClellan differentiator (Parks and McClellan 1972)

- ...

## 5.4.4 Integration

### Analytical Integration

Typical measurement devices for kinematic recordings provide velocity or acceleration. For example, *inertial measurement units (IMUs)* typically provide the linear acceleration of an object.

In order to obtain linear velocity and position from linear acceleration and velocity, respectively, these data have to be integrated:

$$\mathbf{vel}(t) = \mathbf{vel}(t_0) + \int_{t_0}^{t} \mathbf{acc}(t') \, dt' \tag{5.22}$$

$$\mathbf{x}(t) = \mathbf{x}(t_0) + \int_{t_0}^{t} \mathbf{vel}(t'') \, dt'' =$$

$$= \mathbf{x}(t_0) + \mathbf{vel}(t_0) * (t - t_0) + \int_{t_0}^{t} \int_{t_0}^{t''} \mathbf{acc}(t') \, dt' \, dt'' \tag{5.23}$$

If the sensor at $t_0$ has a velocity $\mathbf{v}_0$, the change in position is given by

$$\Delta\mathbf{x}(t) = \mathbf{x}(t) - \mathbf{x}(t_0) = \mathbf{v}_0 * \Delta t + \int_{t_0}^{t} \int_{t_0}^{t''} \mathbf{acc}(t') \, dt' \, dt'' \tag{5.24}$$

**Note:** The acceleration in Eqs. (5.22)–(5.24) is the acceleration with respect to space. In measurements with inertial sensors, however, the acceleration is obtained with respect to the sensor! The mathematical details of the analysis required to compensate for this change of reference frames are explained in Haslwanter (2018).

### Numerical Integration

When working with discrete data the integral can only be determined approximately. Splitting the time between $t_0$ and $t$ into n equal elements with width $\Delta t$ leads to

$$\mathbf{x}(t_n) = \mathbf{x_0} + \Delta\mathbf{x_1} + \Delta\mathbf{x_2} + \cdots + \Delta\mathbf{x_n} \tag{5.25}$$

Measuring the acceleration at discrete times $t_i (i = 0, \ldots, n)$, Eqs. (5.22) and (5.23) have to be replaced with discrete equations:

$$\mathbf{vel}(t_{i+1}) \approx \mathbf{vel}(t_i) + \mathbf{acc}(t_i) * \Delta t \tag{5.26}$$

$$\mathbf{x}(t_{i+1}) \approx \mathbf{x}(t_i) + \mathbf{vel}(t_i) * \Delta t + \frac{\mathbf{acc}(t_i)}{2} * \Delta t^2 \tag{5.27}$$

with the sampling period $\Delta t$.[1] The solution to Eq. (5.25) with the approximation in Eq. (5.27) can be conveniently implemented in Python.

Visually, the integral of a curve is given by the area under the curve. While integration can be performed as IIR-filter (see Exercise 5.9.1), it easier to use the first- and second order approximations provided by the functions `np.cumsum` and `scipy.integrate.cumtrapz` (Fig. 5.19).

---

[1]The numerical stability can be improved by replacing $\mathbf{acc}(t_i)$ with $\mathbf{acc}(t_{i+1})$. This is called the *Euler-Cromer method.*

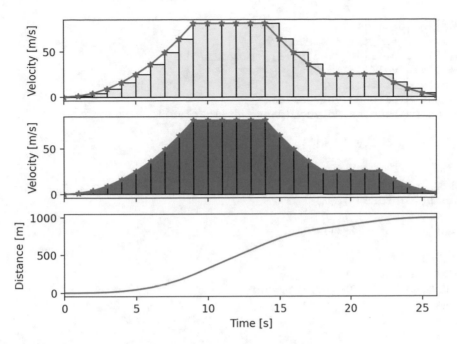

Figure 5.19: Demonstration on how to numerically integrate position data to get velocity. **Top:** the area summed up by `cumsum`. **Middle:** the area summed up by `cumtrapz`. **Bottom:** cumulative integral from `cumtrapz` (from Listing 5.3).

Listing 5.3: integration_demo.py

```python
""" Demonstration on how to numerically integrate a signal """

# Import the standard packages
import numpy as np
import matplotlib.pyplot as plt
from matplotlib import patches
from scipy.integrate import cumtrapz

# Generate velocity data for Fig. 5.19
vel = np.hstack((np.arange(10)**2,
                 np.ones(4) * 9**2,
                 np.arange(9, 4, -1)**2,
                 np.ones(3) * 5**2,
                 np.arange(5, 0, -1)**2))
time = np.arange(len(vel))


## Plot the data
fig, axs = plt.subplots(3, 1, sharex=True)

axs[0].plot(time, vel, '*-')
for ii in range(len(vel)-1):
    ## Corresponding trapezoid corners
    x = [time[ii], time[ii], time[ii+1], time[ii+1]]
    y = [0, vel[ii], vel[ii], 0]
    data = np.column_stack((x,y))
    axs[0].add_patch(patches.Polygon(data, alpha=0.1))
    axs[0].add_patch(patches.Polygon(data, fill=False))
axs[0].set_ylabel('Velocity [m/s]')


axs[1].plot(time, vel, '*-')
```

```
for ii in range(len(vel)-1):
    ## Corresponding trapezoid corners
    x = [time[ii], time[ii], time[ii+1], time[ii+1]]
    y = [0, vel[ii], vel[ii+1], 0]
    data = np.column_stack((x,y))
    axs[1].add_patch(patches.Polygon(data))
    axs[1].add_patch(patches.Polygon(data, fill=False))

axs[1].set_ylabel('Velocity [m/s]')


axs[2].plot(time, np.hstack([0, cumtrapz(vel)]))
axs[2].set_ylabel('Distance [m]')
axs[2].set_xlabel('Time [s]')
axs[2].set_xlim([0, len(vel)-1])

# Save and show the image
out_file = 'numericalIntegrationPy.jpg'
plt.savefig(out_file, dpi=200, quality=90)
print(f'Image saved to {out_file}')

plt.show()
```

## 5.5   Smoothing of Irregularly Sampled Data

Curve smoothing is an important topic, and depending on the requirements different solutions are available. Smoothing options depend for example on how the data have been sampled. If they have been sampled at equal intervals, FIR- or IIR-filters can be used. But if they have been recorded with varying time delays, for example with wireless sensors and occasional samples have been lost, then other approaches are required. Some of these alternatives are listed below.

### 5.5.1   Lowess and Loess Smoothing

Two common approaches for the smoothing of irregularly spaced 1D data are "loess" filters (which stands for *LOcal regrESSion*) and "lowess" filters (*LOcally WEighted Scatterplot Smoothing*). These two are related non-parametric regression methods that combine multiple regression models with a so-called "k-nearest-neighbor-based meta-model". Loess is a generalization of lowess.

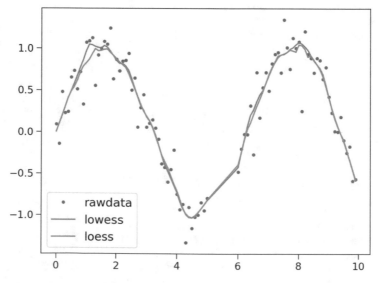

Figure 5.20: *lowess* and *loess* smoothing of noisy data (from Listing 5.4).

In short, one specifies the percentage of adjacent data to be include. For these data, a weighted linear regression is applied. The traditional weight function used for lowess and loess is the tri-cube weight function,

$$w(x) = (1 - |x|^3)^3 \, \mathrm{I}_{[|x|<1]} \tag{5.28}$$

$\mathrm{I}_{[...]}$ is the *indicator function*, indicating the range over which the function argument is True: in this range the function equals to *1*, otherwise it is *0*.

Lowess and loess differ in the model they use for the regression: lowess uses a linear polynomial, while loess uses a quadratic polynomial. Figure 5.20 demonstrates an example of lowess and loess smoothing.

Listing 5.4: lowess.py

```python
""" Lowess and Loess-Smoothing

Note that this script requires the installation of the packages 'loess' and
'statsmodels'!
"""

import numpy as np
import matplotlib.pyplot as plt
from statsmodels.nonparametric.smoothers_lowess import lowess
from loess.loess_1d import loess_1d

np.random.seed(1234)

# Generate some data
x = np.arange(0,10,0.1)
y = np.sin(x) + 0.2 * np.random.randn(len(x))

# Eliminate some, so that we don't have equal sampling distances
cur_ind = np.where( (x>5) & (x<6) )
x_space = np.delete(x, cur_ind)
y_space = np.delete(y, cur_ind)
plt.plot(x_space, y_space, '.', label='rawdata')

# Smooth the data with Lowess, from the package "statsmodels"
smoothed = lowess(y_space, x_space, frac=0.1)
index, data = smoothed.T
plt.plot(index, data, label='lowess')

# Smooth with Loess, from the package "loess"
x_out, y_out, weights = loess_1d(x_space, y_space, frac=0.1)
plt.plot(x_out, y_out, label='loess')
plt.legend()

# Save and show the image
out_file = 'loess.jpg'
plt.savefig(out_file, dpi=200, quality=90)
print(f'Image saved to {out_file}')

plt.show()
```

## 5.5.2 Splines

### Definition

*Interpolations* of data by definition always include the data they are based on (see Sect. 6.4). But this is not always required, and so-called *splines* offer a powerful alternative. Splines can be used not only for interpolation, but also for data smoothing and differentiation.

The ideas of splines have their roots in the aircraft and shipbuilding industries. For example, the British aircraft industry during World War II used to construct templates for airplanes by passing thin wooden strips (called "splines") through points laid out on the floor of a large design loft, a technique borrowed from ship-hull design (Fig. 5.21). In the late 1950s and 60s, the computational use of splines was developed for modeling automobile bodies. In the computer science subfields of computer-aided design and computer graphics, splines are popular because of the simplicity of their construction, their ease and accuracy of evaluation, and their capacity to approximate complex shapes through curve fitting and interactive curve design.

A *spline*-function is nowadays defined as a *piecewise polynomial function* of degree $<k$ in a variable $x$. $k$ is called the "degree of the spline", and $k+1$ the "order of the spline". (I know, it's a bit confusing.)

Figure 5.21: A wooden spline (From the archives of Pearson Scott Foresman, donated to the Wikimedia Foundation.).

## B-Splines

One particularly simple and powerful option to construct smooth, piecewise polynomial 2-D and 3-D trajectories are so-called *B-splines*. The term *B-splines* stands for "basis splines", since any spline function of given degree can be expressed as a linear combination of B-splines of that degree.

For a given trajectory, the *spline-knots* separate the piecewise polynomial parts of the trajectory. If the knots are equidistant, the spline is called *cardinal B-spline*, and the definition of B-splines then becomes remarkably simple: with a B-spline of degree $p$ ($p \in \mathbb{N}_0$), the convolution operator $*$, and the indicator function $\mathbf{b^0} = \mathbf{I}_{[0,1)}$ of the half-open unit interval $[0,1)$, the corresponding cardinal B-splines is given by

$$\mathbf{b^p} := \underbrace{\mathbf{I}_{[0,1)} * \cdots * \mathbf{I}_{[0,1)}}_{p+1-times} \tag{5.29}$$

Note that B-splines have what is called *minimal support*: linear B-splines only have an effect over two adjacent knots, quadratic B-splines over three knots, etc (Fig. 5.22).

Listing 5.5: show_bsplines.py

```
"""Simple generation of B-splines"""

# Import the required packages
import numpy as np
import matplotlib.pyplot as plt

dt = 0.01    # step interval for plotting
t = np.arange(0,1,dt)

# Generate the B-splines, through convolution
```

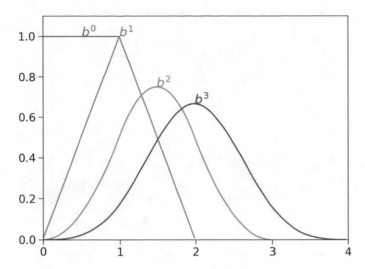

Figure 5.22: The first four B-splines (from Listing 5.5).

```python
ones = np.ones(len(t))
Bsplines = [ones]
for ii in range(3):
    Bsplines.append(np.convolve(Bsplines[-1], ones))
    # Normalize the integral to "1"
    Bsplines[-1] /= np.sum(Bsplines[-1])*dt

# Plot the Bsplines
for spline in Bsplines:
    plt.plot(np.arange(len(spline))*dt, spline)

# Put on labels
plt.text(0.5, 1, '$b^0$', color='C0')
for ii in range(1,4):
    spline = Bsplines[ii]
    loc_max = np.argmax(spline)*dt
    val_max = np.max(spline)
    txt = '$b^' + str(ii) + '$'      # e.g. '$b^1$'
    color = 'C' + str(ii)            # e.g. 'C1'
    plt.text(loc_max, val_max, txt, color=color)

# Format the plot
plt.xlim(0, 4)
plt.xticks(np.arange(5))
plt.ylim(0, 1.1)

# Save and show the image
out_file = 'Bsplines.jpg'
plt.savefig(out_file, dpi=200, quality=90)
print(f'Image saved to {out_file}')

plt.show()
```

A B-spline curve $C(u)$, $u \in [\tau_p, \tau_{n-p-1}[$ of grade $p$ with knot vector $\tau$ and control points $P_i$ $(i = 0, \ldots, n - p - 2)$ (also called *De-Boor-points*) is given by

$$C(u) = \sum_{i=0}^{n-p-2} P_i \, B_{i,p}(u) \tag{5.30}$$

where $B_{i,p}$ are the $b^p$ in Fig. 5.22, shifted over to point $i$. In one dimension the construction of a linear spline by three control points can be easily visualized (Fig. 5.23):

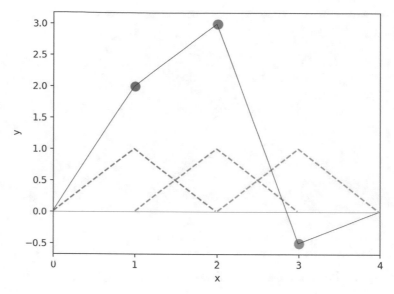

Figure 5.23: Explanation of B-spline curves. This example shows three control points, and the three corresponding linear B-splines. Multiplying each point with the corresponding (dotted) B-spline, and summing up the results, gives the solid line. Note that each point contributes only to a limited interval of the total curve.

A different formulation, which also works for knots that are *not* equidistant, has been given by DeBoor. That formula involves recursive equations and is implemented in `scipy.inter polate.Bspline`. Since by convention the first spline, $b^0$ in Fig. 5.22, is called "B-spline of order 1", B-splines "of oder n" contain polynomials of degree $n-1$.

### 5.5.3   Kernel Density Estimation

In many applications discrete data samples are given, but smooth "density distributions" of the data are desired. For example, with one-dimensional data the *frequency* of events is often indicated with histograms, and with two-dimensional data as scatter-plots. To obtain smooth probability density functions from such data, the *Kernel Density Estimation (KDE)* technique can be used.

Figure 5.25 gives an example for a one-dimensional data set. For a KDE of 1D data, each sample is multiplied with the Gaussian function

$$g(x) = \frac{1}{\sigma\sqrt{2\pi}}e^{-\frac{1}{2}\left(\frac{x-\mu}{\sigma}\right)^2}. \tag{5.31}$$

and the resulting curves are then summed up. A combination of a histogram, the original data, and the corresponding KDE, is provided by the function `distplot` from the package `seaborn`. *Seaborn* offers numerous tools and functions for the display of statistical data, including 1D- and 2D-KDEs.

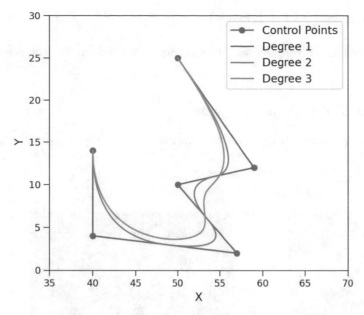

Figure 5.24: Control points and corresponding B-splines with polynomials of varying degrees (from CQ `bspline_demo.py`, Appendix A).

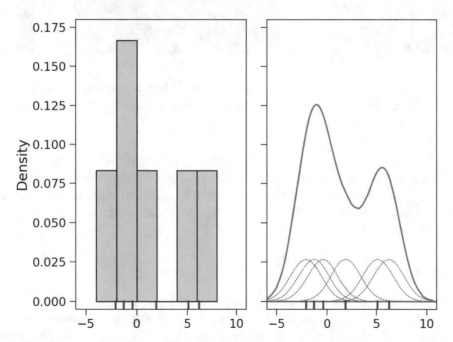

Figure 5.25: **Left:** Histogram. **Right:** Corresponding *Kernel Density Estimation (KDE)*. The ticks on the x-axis indicate individual events, and the thin dotted lines the corresponding Gaussians. The thick solid line is the sum of the thin dotted lines, and provides a continuous "density estimate" for the event rate.

## 5.6    Filtering Images (2D Filters)

Filtering also works in more than 2 dimensions. In this Section we describe the format and the filtering of 2D images.

### 5.6.1    Representation of Grayscale Images

The simplest image type is a grayscale image: there each pixel is given a gray-level corresponding to its brightness (Fig. 5.26)

$$0 < gray\_level < g_{max}$$

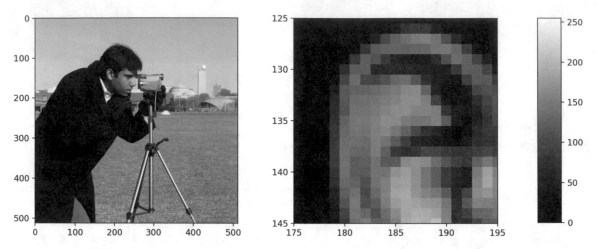

Figure 5.26: Images consist of individual pixels, each of which has a constant gray-level. The middle image is zoomed in on the ear of the photographer, and the scale on the right side shows how pixel brightness corresponds to gray-level.

For many images 8-bit gray-level resolution is sufficient, providing $2^8 = 256$ gray levels. Note that a higher image depth only makes sense if the sensing devices used have measured differences at a finer level! Also keep in mind that a Python `float` typically uses 64 bit. Since this requires 8 times as much memory as unsigned 8-bits, one should only convert the data to `float` when this is really needed.

### 5.6.2    Color Images

A color image is nothing more than a stack of three grayscale images: one representing the "red" channel, one the "green" channel, and one the "blue" channel. If the three colors are stacked in this sequence, the image is referred to as an *RGB-image*. For example, take the astronaut image from `scikit-image` (Fig. 5.27):

```
import skimage as ski
img = ski.data.astronaut()
img.shape
>> (512, 512, 3)
```

The image has 512 pixels horizontally, 512 pixels vertically, and 3 color layers.

### 5.6.3    Image Transparency

Occasionally colored images contain an additional fourth layer, called "alpha layer": that layer specifies the amount of transparency of that image. The technique is called "alpha blending" or "alpha compositing". An example is given in Fig. 5.28.

Figure 5.27: RGB-image, from `scikit-image`.

### 5.6.4 2D-Filtering

Filtering similar to FIR in 1D can also be performed on 2D signals, such as images. Instead of a 1D input and a weight *vector*, we now have a 2D input and a weight *matrix* (Fig. 5.29). The area around a pixel that influences the filter output is sometimes called "structural element" (SE) (Fig. 5.29). It does not have to be square, but can also be rectangular, circular, or elliptical. The output is still obtained by summing up the weighted input:

$$y(n,m) = \sum_{i=-k}^{k} \sum_{j=-l}^{l} w_{ij} * x(n+i, m+j) \tag{5.32}$$

The moving window interpretation still holds, except now

- the window and data extend in both dimensions, and

- the sequence of the kernel indices is not inverted. So this corresponds more to a 2D cross-correlation than to a 2D convolution, and the easiest way to apply a filter-kernel to an image is with the command `scipy.ndimage.correlate`. (See also Sect. 6.2.6.)

For example, in order to blur an image, enhance horizontal lines, or enhance vertical lines, one can use the following lines of code (Fig. 5.30):

```
Listing 5.6: filter_images.py

""" Demonstration on how to filter images """

# author: Thomas Haslwanter
# date:    Oct-2019

# Import the standard packages
import numpy as np
import matplotlib.pyplot as plt
import os

# For the image filtering
from scipy import ndimage
```

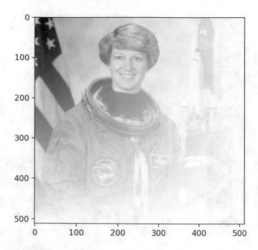

Figure 5.28: A transparency layer has been added to the image, so that it fades form top-right to bottom-left (from CQ `fading_astronout.py`, Appendix A).

```
# Import formatting commands
from utilities.my_style import set_fonts, show_data

# Get the data
import skimage as ski
img = ski.data.camera()
img_f = np.array(img, dtype=float)    # for the filtering, the data must not be
    uint

# Make the filters: one for averaging, and two for edge detection
Filters = []
Filters.append(np.ones((11,11))/121)
Filters.append(np.array([np.ones(11),np.zeros(11),-1*np.ones(11)]))
Filters.append(Filters[-1].T)

# Filter the images
filtered = []
for filt in Filters:
    filtered.append( ndimage.correlate(img_f, filt) )

# Make the plots
fig, axs = plt.subplots(3,2, figsize=(6,8))
plt.gray()

axs[0,0].imshow(img)
axs[0,1].imshow(filtered[0])
axs[1,0].imshow(filtered[1])
axs[1,1].imshow(filtered[2])
axs[2,0].imshow(filtered[1]>125)
axs[2,1].imshow(filtered[2]>125)

# Remove the ticks and labels
for axis in axs.ravel():
    axis.axes.get_xaxis().set_visible(False)
    axis.axes.get_yaxis().set_visible(False)

# Reduce the space between the plots
plt.tight_layout()

# Save and show the figure
out_file = 'filter_demo.jpg'
show_data(out_file)
```

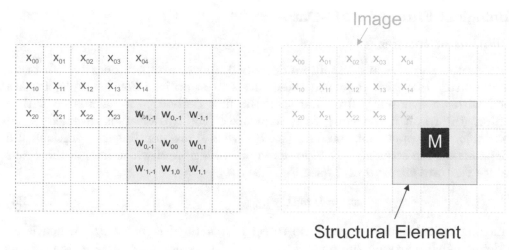

Figure 5.29: **Left:** the $X_{ij}$ indicate the gray-level values at the pixel in row $i$ and column $j$. The two-dimensional filter with a weight-matrix $\mathbf{W}$ is indicated by the weight-values $W_{kl}$. **Right:** the *Structural Element* is defined by the shape of the weight-matrix. Note that a structural element does not have to be a square; it can also be a circle, a rectangle, etc. The output value at pixel $M$ is obtained by multiplication of the weight-values with the underlying image values, and summing up of the result.

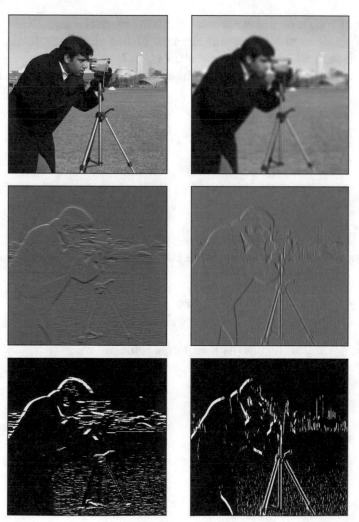

Figure 5.30: **Top row:** Original image (left), and blurred version (right). **Middle row:** Horizontal (left), and vertical (right) edges enhanced. **Bottom row:** Same data as middle row, but black-and-white converted, with a threshold of 125 (from Listing 5.6).

### 5.6.5   Morphological Filters for 2D-Data

**Erosion and Dilation of Images**

Linear filters, such as the the 1D- and 2D-filters described above, are commutative. In other words, it does not matter in which sequence these filters are applied. For example, if data are first smoothed with an averaging filter and then the first derivative is calculated with a Savitzky–Golay filter, the result is the same if the sequence of the two filters is reversed.

In contrast, *morphological* operations on data (see Sect. 5.2.4) are non-linear operations and the final result depends on the sequence. Take for example a black-and-white image, the pixels of which have values that are either *0* or *1* (see Fig. 5.31):

$$x_{ij} = 0 \ or \ 1, \forall i, j \tag{5.33}$$

To define a morphological operation we have to set a *structural element SE*, which is an area with a well defined shape around the point *M*. In Fig. 5.31 we use a square $3 \times 3$-matrix as SE.

The definition of the action *erosion E* is:

$$E(M) = \begin{cases} 0, & \text{if } \sum_{i=0}^{n} \sum_{j=0}^{m} se_{ij} < 9, \text{with } se_{ij} \in SE(M) \\ 1, & \text{else} \end{cases} \tag{5.34}$$

with *n* and *m* the height and width of the structural element, respectively.

In words, if any of the pixels in the structural element M has the value *0*, the erosion sets the value of M, a specific pixel in SE, to *0*. Otherwise *E(M) = 1*.

And for the *dilation* D it holds that if any value in SE is 1, the dilation of M, *D(M)*, is set to 1.

$$D(M) = \begin{cases} 1, & \text{if } \sum_{i=0}^{n} \sum_{j=0}^{m} se_{ij} \geq 1, \text{with } se_{ij} \in SE(M) \\ 0, & \text{else} \end{cases} \tag{5.35}$$

**Opening and Closing of Images**

Depending on the sequence, a dilation and an erosion can be combined to either an *opening* or a *closing* action. These are defined as:

$$\begin{aligned} opening &= dilation \circ erosion \\ closing &= erosion \circ dilation \end{aligned} \tag{5.36}$$

where ∘ indicates "after" (i.e. the same rules as for matrix multiplications).

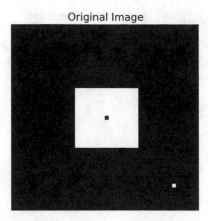

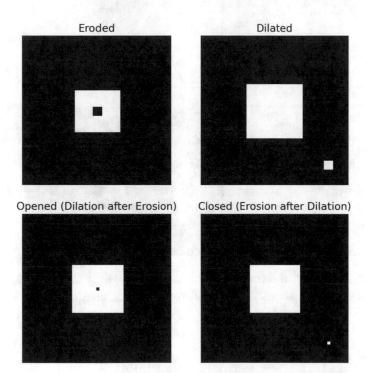

Figure 5.31: Compositions of dilation and erosion: opening and closing of images. While "opening" an image removes small white spots on a dark background (bottom left), "closing" an image removes black spots on a white background (bottom right) (from CQ `morphology.py`, Appendix A).

## Practical Example

To give an example how much can be achieved with just a few lines of image processing code, the Listing below shows how to extract the pupil edge from the image of an eye (Fig. 5.32).

Listing 5.7: find_pupil.py

```
""" Find the pupil-edge in an image of the eye, using sckit-image """

# Import standard modules
import os
import numpy as np
import matplotlib.pyplot as plt
```

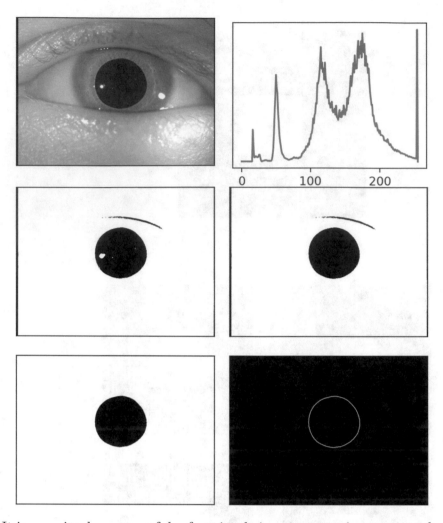

Figure 5.32: It is amazing how powerful a few simple image processing commands can be! From top left: original image, histogram, thresholding, filling-of-holes, closing, and edge detection (from Listing 5.7).

```python
# Modules for image processing
from scipy import ndimage
from skimage import morphology
from skimage import filters

def show_me(data):
    '''Show image data in graylevel'''
    plt.imshow(data, cmap='gray')
    plt.show()

if __name__ == '__main__':

    # Get the data ...
    data_dir = r'..\..\Data'
    file_name = 'eye.bmp'
    in_file = os.path.join(data_dir, file_name)
    data = plt.imread(in_file)
    show_me(data)

    # Calculate and show the histogram
    histo = np.histogram(data, bins=np.arange(0,256))
    plt.plot(histo[1][:-1], histo[0])
    plt.show()
```

```
# Convert to black-and-white
# for convenience, I choose the threshold automatically
bw = data>80
show_me(bw)

# Fill the holes
filled = np.invert(ndimage.binary_fill_holes(np.invert(bw)))
show_me(filled)

# 'Close' the image, with a disk (radius 10 pixels) as structural element
se = morphology.disk(10)
closed = morphology.closing(filled, se)
show_me(closed)

# Edge detection
edges = filters.sobel(closed)
show_me(edges)
```

## 5.7  Exercises

1. **Integration with an IIR-filter**

   - Show how the function `cumsum` can be implemented as an IIR-filter.

   - Show how the function `cumtrapz` can be implemented as an IIR-filter.

   - Apply the integration functions to

     (a) the vector 1:5, and

     (b) to one cycle of a sine-wave.

2. **Smoothing and Differentiation of Noisy Data**

   - Create two cycles of a noisy sine wave, with an amplitude of 1, a frequency of 0.3 Hz, a sampling rate of 100 Hz, and a Gaussian random noise with a standard deviation of 0.5.

   - Using a Savitzky–Golay filter, smooth the noisy data, and superpose the smoothed data over the original ones.

   - Using a Savitzky–Golay filter, calculate the first derivative using a $2^{nd}$ order polynomial. Try different window sizes, and superpose the results in a plot with the ideal sinusoid (i.e. the sine-wave without the noise). Make sure that the axes of the plot are correctly labeled.

3. **Analyze EMG-data**
   Electro-Myography (EMG)-data are some of the most common signals in movement analysis. But sometimes the data analysis is not that simple. For example, data can be superposed by spurious drifts. And short drops in EMG activity can obscure extended periods of muscle contractions.
   The data in `Shimmer3_EMG_Calibrated.csv` have been taken from https://www.shimmersensing.com/support/sample-data/, where paradigm and data description are presented in detail.
   Write a function that does the following:

- Import the EMG data from the data file `Shimmer3_EMG_Calibrated.csv`
- Remove the offset and drift of the EMG-recording
- Rectify the data, and smooth them to produce a rough envelope of the signal indicating the activation of the muscle
- Find the start- and end-points of muscle activity
- Eliminate artifacts
- Show the filtered data, as well as the detected start- and end-times of the muscle contractions
- Calculate and display the mean contraction time.

4. **Band-pass Filter**
Generate a dummy data set consisting of the sum of three sine-waves, with frequencies of *[2, 30, 400]* Hz, and amplitudes of *[0.5, 1, 0.1]*. The signal should have a sampling rate of *5* kHz, and a duration of *2* s. Now generate a 3rd-order *Butterworth band-pass filter*, for the frequency band between *10* and *100* Hz. Apply that filter to your data: you should obtain a pure sine-wave with *30* Hz.

5. **Exponential Averaging Filter**
Apply the *exponential averaging filter* to a step-input. Plot the response for different values of $\alpha$, to show why this filter is in some fields of research referred to as "leaky integrator".

# References

Haslwanter, T. (2018). *3D kinematics*. Berlin: Springer.

Lanczos, C. (1961). *Applied analysis*. Englewood Cliff, N.J.: Prentice Hall.

Madden, H. H. (1978). Comments on Savitzky-Golay convolution method for least-squares fit smoothing and differentiation of digital data. *Analytical Chemistry, 50*, 1383–1386.

Parks, T. W., & McClellan, J. H. (1972). Chebyshev approximation for nonrecursive digital filters with linear phase. *IEEE Transactions on Circuit Theory, CT-19*(2), 189–194.

Press, W., Teukolsky, S., Vetterling, W., & Flannery, S. (2007). *Numerical recipes in C* (3rd ed.). Cambridge: Cambridge University Press.

Savitzky, A., & Golay, M. J. (1964). Smoothing and differentiation of data by simplified least squares procedures. *Analytical Chemistry, 36*, 1627–1639.

Smith, III, J. O. (2007). *Introduction to digital filters: With audio applications*. Stanford: W3K Publishing.

Wiener, N. (1942). *The interpolation, extrapolation and smoothing of stationary time series*. NDRC report. New York, Cambridge: Wiley.

# Chapter 6

# Event- and Feature-Finding

Often it is necessary to find specific locations in a stream of data. For example, one might want to find out when a signal passes a given threshold (Fig. 6.1). Or when analyzing movement signals, one might want to find at which point the movement starts, and the point at which it ends. If the data is a time series these locations are often referred to as *events*.

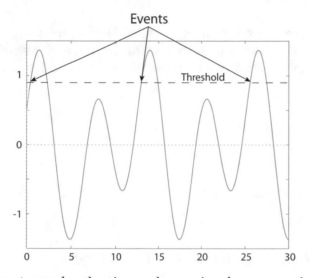

Figure 6.1: *Events* can be the times when a signal passes a given *Threshold*.

## 6.1 Finding Simple Features

There are two methods to access single events in Python:

- logical indexing

- the functions `np.where` and `np.nonzero`

We will illustrate these methods with an example. Let us take the values

```
time = np.arange(3, 9)
signal = time/10          # just to have some values
```

To check whether `time` is larger than 5 one can use a logical comparison:

```
is_late = time > 5
```

© The Author(s), under exclusive license to Springer Nature Switzerland AG 2021
T. Haslwanter, *An Introduction to Hands-on Signal Analysis with Python*,
https://doi.org/10.1007/978-3-030-57903-6_6

The result of the comparison, which is now stored in the variable is_late, is a Boolean vector

```
>> array([False, False, False,  True,  True,  True])
```

Boolean arrays can be inverted with the ∼ operator:

```
    is_early = ~is_late
>> array([True,  True,  True, False, False, False])
```

To find the corresponding "late" data in signal, the easiest (and often most efficient) method is

```
    signal[is_late]
```

or equivalently

```
    signal[time>5]
```

This method of selecting elements from an array is called "logical indexing", since a Boolean array the same size as the indexed array is used to address the True elements.

If the explicit indices of True events are required, then the function np.where can be used[1]:

```
    late_indices = np.where(is_late)[0]
```

or equivalently

```
    late_indices = np.where(time > 5)[0]
```

The [0] is required since np.where returns lists.

## 6.1.1   Example 1: Find Large Signal Values in 1D-Data

Listing 6.1: events.py

```
1  """ Show logical indexing """
2
3  # Import the standard packages
4  import numpy as np
5  import matplotlib.pyplot as plt
6
7  # Create a sine-wave
8  dt = 0.1
9  duration = 20
10 time = np.arange(0,duration,dt)
11 data = np.sin(time)
12
13 # Set a threshold
14 threshold = 0.7
15
16 # Find the (binary) indices of all data above that threshold
17 is_large = data > threshold
18
19 # For plotting of "large" data, set all "not large" data to "np.nan"
```

---

[1]Here the simpler function np.nonzero would be equivalent; but personally, I prefer the "sound" of np.where.

```
20 # Note that I explicitly copy the data!
21 large_data = data.copy()
22 large_data[~is_large] = np.nan
23
24 # Plot the data
25 fig, axs = plt.subplots(3,1)
26
27 axs[0].plot(time, data)
28 axs[0].plot(time, large_data, lw=3)
29 axs[1].plot(time[is_large], data[is_large], '*-')
30 axs[2].plot(data[is_large])
31
32 # Format the plot
33 axs[0].set_ylabel('All data')
34 axs[0].axhline(threshold, ls='dotted')
35 axs[0].margins(x=0)
36 axs[0].set_xticklabels([])
37
38 axs[1].set_ylabel('Large data')
39 axs[1].set_xlabel('Time [s]')
40 axs[1].margins(x=0)
41 axs[1].set_xlim(0, duration)
42 axs[1].set_ylim(-1.05, 1.05)
43
44 axs[2].set_ylabel('Large data only')
45 axs[2].set_xlabel('Points only')
46
47 # Group the top two axes, since they have the same x-scale
48 axs[0].set_position([0.125, 0.75, 0.775, 0.227])
49 axs[1].set_position([0.125, 0.50, 0.775, 0.227])
50 axs[2].set_position([0.125, 0.09, 0.775, 0.227])
51
52 # Save and show the figure
53 out_file = 'FindingEvents.jpg'
54 plt.savefig(out_file, dpi=200, quality=90)
55 print(f'Image saved to {out_file}')
56
57 plt.show()
```

This code produces Fig. 6.2. Some segments of this code are noteworthy:

- **21/22 & 28:** To plot only the "large" data, all "not large" data are set to `np.nan`. This can only be done for `float` arrays, since `np.nan` has the size `float`. Also, since Python copies by reference, not by value, it is important to explicitly `copy` the array in line 21! Otherwise the following can happen:

```
x = [1, 2]
y = x
y[1] = 10
print(x)
>> [1, 10]
```

This is well described in more detail in https://realpython.com/copying-python-objects/. (*Real Python* is in general a site with lots of valuable information on Python programming.) Since other programming languages such as Matlab copy by value, it is important to understand this behavior of Python in order to avoid mistakes that are difficult to find.

- **29:** To plot "large" data at the correct locations, the corresponding time-values `time[is_large]` are used in this plot command. Note that connecting those points with a solid

108 CHAPTER 6. EVENT- AND FEATURE-FINDING

line leads to the artifacts in the middle axis in Fig. 6.2, if the intermediate points are *not* set to np.nan. (The correct way to do it is shown in lines 21/22 & 28.)

- If only the "large" values are selected for the plot, all other values are discarded (bottom axis in Fig. 6.2).

- **34:** The axes methods axhline / axvline are convenient for drawing coordinate axes or similar lines that extend over the whole axis. These commands are also accessible from matplotlib.pyplot (plt).

- **35:** The axes-method margins(x=..., y=...) is useful for eliminating the default plot margins in *matplotlib* figures. (Using this command, the xlim in the top axes is determined exactly by the range of the x-values; in contrast, a small x-margin is use in the bottom axis.)

- **36:** Shows how ticklabels can be eliminated.

- **48–50:** To indicate that the top two axes share the same x-values, they are manually positioned closely together. The bottom axis is moved down a bit.

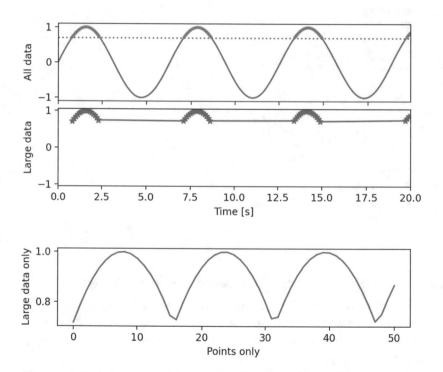

Figure 6.2: Selection of large signal values (from Listing 6.1).

## Combine Two Logical Conditions

If we want to eliminate all large signals that occur in the first 10 s, we could add the following code:

```
#First find the late section ...
    time_threshold = 10
    is_late = time > time_threshold

# ... and combine it with the other criterion with the bit-wise AND operator '&'
is_late_and_large = is_late & is_large
signal_trimmed = np.delete(signal, np.where(is_late_and_large)[0])
```

## 6.1.2 Example 2: Find Start and End of a Movement

Real measurement data will always contain noise (and usually a lot more than you want). To remove noise-effects from the analysis it is often helpful to go from continuous data to binary `True/False` representations.

For example, finding periods of movement in position recordings (e.g. Fig. 6.3) could be achieved with the following analysis steps:

1. Find a threshold (i.e. a signal value that has to be reached in order to be part of the "movement" data set).

2. For each data-point, calculate if the signal is above the threshold (`True/False`).
   At this stage, we have gone from a noisy, real-valued signal, to a discrete, binary signal. (Tip: in most cases it will be better so set the threshold manually, because the the noise level may vary from recording to recording.)

3. Now the start of our feature can be found easily: it is the point where `np.diff(binary_signal*1)` equals 1. (The multiplication of `binary_signal` with 1 is required to convert the binary `True/False` signal to a numerical `1/0` signal.) Similarly, the end of our feature can be found by checking where `np.diff(binary_signal*1)` equals −1.

The essence of event-finding is contained elegantly in this example (Fig. 6.3), which uses real eye-position recordings.

**Listing 6.2: event_detection.py**

```python
""" Show how events can be elegantly detected using binary indexing """

# Import the standard packages
import numpy as np
import matplotlib.pyplot as plt
import matplotlib as mpl
import os
from scipy import signal, io
from pprint import pprint

# Get eye positions, sampled with 100 Hz
data_dir = r'..\..\data'
file_name = 'HorPos.mat'
in_file = os.path.join(data_dir, file_name)

rate = 100
data = io.loadmat(in_file)
hor_pos = data['HorPos'][:,0]

# General layout of the plot
fig, axs = plt.subplots(3,2)
mpl.rc('lines', lw=0.8)

# Select an interesting domain
my_domain = range(9000, 13500, 1)
axs[0,0].plot(hor_pos[my_domain])
axs[0,0].set_ylabel('Position')
axs[0,0].tick_params(labelbottom=False)
axs[0,0].margins(x=0)

# Plot the absolute eye velocity
eye_vel = signal.savgol_filter(hor_pos, window_length=71, polyorder=3,
        deriv=1, delta=1/rate)
```

```python
eye_vel_abs = np.abs(eye_vel)
axs[1,0].plot(eye_vel_abs[my_domain])

# Set a default threshold, in case the threshold is not determined interactively
threshold = 6.3
axs[1,0].axhline(threshold, color='C1')
axs[1,0].tick_params(labelbottom=False)
axs[1,0].margins(x=0)
axs[1,0].set_ylabel('Velocity')

#To find the threshold interactively, use the following lines
# set(gcf, 'Name', 'Select the threshold:')
# selectedPoint = ginput(1);
# threshold = selectedPoint(2); % I only want the y-value
# set(gcf, 'Name', '');

# Plot3: show where the absolute velocity exceeds the threshold
is_fast = eye_vel_abs > threshold
axs[2,0].plot(is_fast[my_domain], 'o-', ms=2)
axs[2,0].set_ylabel('Above threshold')
axs[2,0].margins(x=0)

# Plot4: as Plot3, but zoomed in
close_domain = range(9900, 10600, 1)
axs[0,1].plot(is_fast[close_domain], 'o-', ms=2)
axs[0,1].set_ylabel('Above threshold')
axs[0,1].tick_params(labelbottom=False)
axs[0,1].margins(x=0)

# Plot5: Find the start and end of each movement
start_stop = np.diff(is_fast*1)        # "*1": to convert boolean signal to
    numerical
axs[1,1].plot(start_stop[close_domain])
axs[1,1].set_ylabel('Start / Stop')
axs[1,1].margins(x=0)

axs[2,1].axis('off')
plt.tight_layout()

# Save and show the figure
out_file = 'event_detection.jpg'
plt.savefig(out_file, dpi=200, quality=90)
print(f'Image saved to {out_file}')

plt.show()

# Find the start and end times for all movements (in sec)
movement = {}
movement['StartTimes'] = np.where(start_stop ==  1)[0]/rate
movement['EndTimes']   = np.where(start_stop == -1)[0]/rate
pprint(movement)
```

### 6.1.3   Example 3: Find Bright Pixels in Grayscale Image

Many methods to find features in one-dimensional signals can also be applied to 2D signals, for example in the field of image processing. For example, to find the bright pixels in a grayscale image all we need to do is apply a threshold (Fig. 6.4):

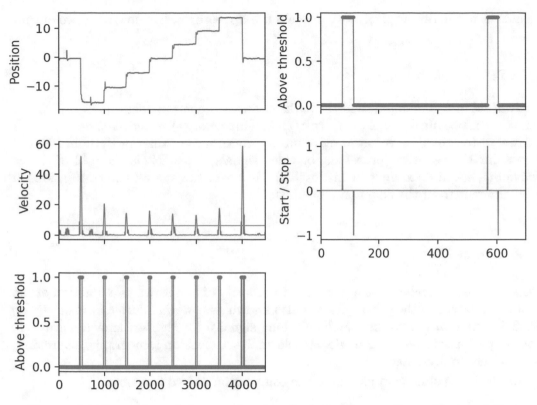

Figure 6.3: Detection of the start and end of movements (from Listing 6.2).

```python
import skimage as ski
img = ski.data.camera()

bright_threshold = 120
is_bright = img > bright_threshold

plt.imshow(is_bright)
```

Figure 6.4: **Left:** grayscale image, with each pixel represented by a `uint8` value. **Right:** corresponding binary image, showing the areas above the threshold in white and the rest in black.

To get the indices of the bright pixels we can again use `np.where`, but now with two output variables:

```
rows, cols = np.where(is_bright)
```

For more advanced image manipulation the package *scikit-image* provides a powerful collection of image processing tools (https://scikit-image.org/). *scikit-image* builds on `scipy.ndimage` to provide a versatile set of image processing routines in Python. If speed is important (e.g. for the real-time processing of video signals), *openCV* is the tool of choice (https://opencv.org/, see also example in Listing 3.2). However, the syntax and conventions of *openCV* can be different from the common Python style.

## 6.2   Cross-Correlation

In signal processing, *cross-correlation* is a measure of similarity of two series as a function of the displacement of one relative to the other.[2] This is also known as a *sliding dot product* or *sliding inner-product*. It is commonly used for searching a long signal for a shorter, known feature. It has applications in pattern recognition, single particle analysis, electron tomography, averaging, cryptoanalysis, and neurophysiology.

The cross-correlation is similar in nature to the convolution of two functions.

### 6.2.1   Comparing Signals

Let us consider how we might assess the similarity of two signals, which we call here *signal* and *feature* (see Fig. 6.5). To find similarities we need some kind of "similarity function" such that the function has a maximum when the feature matches the signal, and that decreases as the difference between *signal* and *feature* increases.

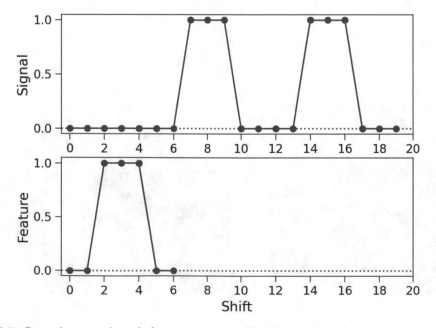

Figure 6.5: Sample *signal* and *feature* to visualize the principle of cross-correlation.

---

[2]Confusingly, in time series analysis and statistics a slightly different definition of "cross-correlation" is used. There the mean of each signal gets subtracted, and the signal normalized by division through its standard deviation (see "The analysis of time series" by C. Chatfield and H. Xing, CRC Press 2019).

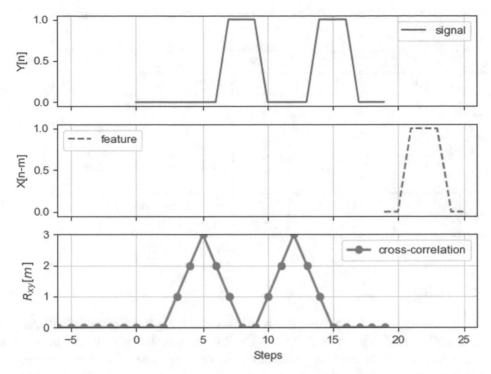

Figure 6.6: Output of the program `corr_vis.py`, to interactively visualize the principle of cross-correlation (from CQ `corr_vis.py`, Appendix A).

It can be shown that the dot-product satisfies both of these properties. Thus, all we need to do to compare part of the *signal* with the *features* is to multiply that part of the *signal* with the *feature*! If we want to find out how much the feature needs to be shifted to match the signal, we calculate the similarity for different relative shifts and choose the shift with the maximum similarity.

In the following we will demonstrate the principles of cross-correlation based on the *signal* and the *feature* shown in Fig. 6.5. The following characteristics can be readily seen:

- In the starting position shown in Fig. 6.5 the dot product between *signal* and *feature* is zero.

- A shift of the *feature* by 5 or 12 steps produces maximum overlap.

- With maximum overlap the dot product between *signal* and *feature* is 3.

- The *feature* can be shifted by six points to the left, before it goes "out of range".

For multiplications with elements outside the given signal/feature-range (e.g. point 20 in Fig. 6.6), the corresponding missing data are commonly replaced by zeros.

That attached program `corr_vis.py` (see the link below) allows an interactive exploration of the sliding dot-product to produce the cross-correlation of *signal* and *feature*. The result is shown in Fig. 6.6.

To demonstrate the principle of cross-correlation, Table 6.1 explicitly goes through the calculation for the signals $sig\_1 = [2, 1, 4, 3]$ and $sig\_2 = [1, 3, 2]$:

The same result is obtained with the *numpy* command `np.correlate(sig_1, sig_2, mode='full')`. (The last option specifies how missing values at the beginning and at the end should be handled; in Table 6.1 they have been set to 0.)

To summarize, cross-correlation provided two pieces of information:

- *How similar* signal and feature are (through the maximum of the cross-correlation).

- *Where* the similarity occurs (through the location of the maximum).

Table 6.1: Sample calculation of cross-correlation. Missing elements at the beginning and at the end of *sig_1* are here set to 0

| 0 | 0 | 2 | 1 | 4 | 3 | 0 | 0 | |
|---|---|---|---|---|---|---|---|---|
| 1 | 3 | 2 | | | | | | $\rightarrow 2*2 = 4$ |
| | 1 | 3 | 2 | | | | | $\rightarrow 2*3 + 1*2 = 8$ |
| | | 1 | 3 | 2 | | | | $\rightarrow 2*1 + 1*3 + 4*2$ $= 13$ |
| | | | 1 | 3 | 2 | | | $\rightarrow 1*1 + 4*3 + 3*2$ $= 19$ |
| | | | | 1 | 3 | 2 | | $\rightarrow 4*1 + 3*3 = 13$ |
| | | | | | 1 | 3 | 2 | $\rightarrow 3*1 = 3$ |

### 6.2.2　Auto-correlation

If the two signals being compared are the same the result is called the *auto-correlation function*. Naturally, the auto-correlation function is not used to find events. Rather it can be used to detect periodicity in a signal which may be impossible to see otherwise. After accounting for the mean offset, the auto-correlation is also used to detect the energy in the signal since the energy in a harmonic signal is proportional to the square of the amplitude.

For example, the commands

```
auto_corr = np.correlate(signal, signal, 'full')
plt.plot(auto_corr)
```

produce the auto-correlation function of the signal in Fig. 6.5. The result is shown in Fig. 6.7. The option `full` for the function `np.correlate` returns the cross correlation at each point of overlap, with an output shape of $(n + m - 1, )$, where $n$ is the length of the first vector and $m$ that of the second one. Note that the auto-correlation is by definition always symmetrical, and has the maximum value for zero shift.

In an auto-correlation, which is the cross-correlation of a signal with itself, there will always be a peak at a lag of zero, and its size will be the signal energy.[3]

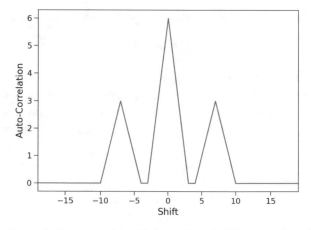

Figure 6.7: Auto-correlation of the *signal* in Fig. 6.5 (top). The results show that this signal has a periodicity with a shift of seven points.

---

[3]Taken from https://en.wikipedia.org/wiki/Cross-correlation.

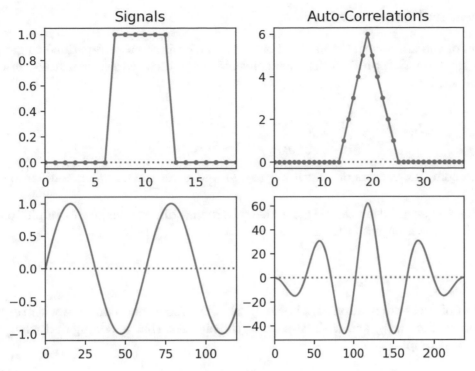

Figure 6.8: Examples of the auto-correlation function of some simple functions. Note that a constant offset leads to a triangle in the auto-correlation (top row).

**Auto-correlation Functions for Simple Signals**

### 6.2.3 Normalization

Sometimes it is desirable to compare the shape of signals, regardless of their amplitude. For example, myo-electric prostheses use EMG-signals for their control; in that application we would like the prosthesis to behave in a repeatable manner, independently of the current quality of the electrode contacts.

In order to evaluate the shape of a signal regardless of its overall duration, amplitude, or offset, we have to *normalize* the signal. Thereby the normalization has to account for three aspects of the signal:

**Offset** To eliminate effects from a constant offset we can subtract the mean of the signal. This avoids the triangular artifacts that arises from a constant offset (Fig. 6.8, top row). Or we can subtract the smallest value of the signal, thus ensuring that the output of the cross-correlation is always positive.

**Duration** To ensure that two signals have the same length we can interpolate them to a fixed number of data points (see Sect. 6.4).

**Amplitude** The most common way to normalize the amplitude of a signal for cross-correlation is such that the *maximum value of the auto-correlation function = 1*. This can be achieved with

$$sig_{normalized} = \frac{sig_{raw}}{\sqrt{\max(autocorr(sig_{raw}))}} \tag{6.1}$$

A signal with the offset eliminated by subtracting the smallest value and the amplitude normalized with Eq. (6.1) has an amplitude between 0 and 1: the amplitude is exactly 1 if the two signals match, and 0 if there is no match at all. This makes it easy to interpret the resulting correlation value.

## 6.2.4   Mathematical Implementation

For the calculation of the cross-correlation the shorter `feature` is often zero-padded to get the same length $n$ as the longer `signal`. With this convention the cross-correlation function of $x$ and $y$ can be obtained by

$$R_{xy}(m) = \begin{cases} \sum_{i=0}^{n-m-1} x_{m+i}y_i^*, & 0 \le m < n-1 \\ R_{yx}^*(-m) & , 1-n < m < 0 \end{cases} \tag{6.2}$$

where $x$ and $y$ are complex vectors of length $n$ ($n > 1$) and $y^*$ is the complex conjugate of $y$ (see Eq. (1.4)).

The concept of cross-correlation can be generalized to higher dimensions, for example to two dimensions to find a known object in an image.

**Notes:**

- The definition of cross-correlation $R_{xy}$ in Eq. (6.2) is not fixed, and can vary with respect to the sequence of $x$ and $y$ and with respect to the element that is conjugated when the input values are complex.

- To optimize speed the cross correlation is often implemented not directly using Eq. (6.2), but rather via the Fast Fourier Transform (FFT). (See Sect. 9.4.)

## 6.2.5   Features of Cross-Correlation Functions

**Length of Cross Correlation Function**

If the signal has a length of $n$ points and the pattern a length of $m$ points, then the cross-correlation has a length of $n + m - 1$ points. For example, the auto-correlation function of a signal with 10 points has a length of 19 points. Note that sometimes the shorter vector is zero-padded to the length of the longer vector, leading to an output with a length of $2*n - 1$.

**Maximum of Cross Correlation Function**

If the signal is multiplied by a factor of $a$, and the feature multiplied by a factor or $b$, then the maximum of the cross-correlation function increases by a factor $a*b$. As a result, if a signal increases by a factor of $a$, the maximum of the auto-correlation function increases by a factor of $a^2$.

## 6.2.6   Cross-Correlation and Convolution

There is a close relationship between cross-correlation, convolution, and FIR-filters.

- As already discussed in Sect. 5.2.2, a convolution of a signal $\mathbf{x}$ with a kernel $\mathbf{w}$ is equivalent to applying an FIR-filter with weights $\mathbf{w}$ to a signal $\mathbf{x}$.

- Apart from a trivial shift in the index, the commands `np.correlate(x,y,'full')` and `np.convolve(x, y[::-1])` produce the same output (see Fig. 6.9).

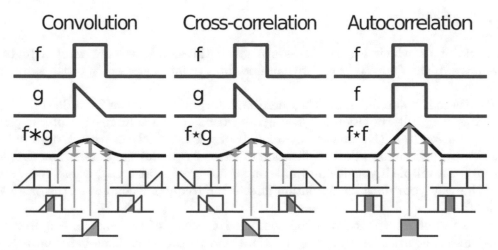

Figure 6.9: Visual comparison of convolution, cross-correlation and auto-correlation. For the operations involving function f, and assuming the height of f is 1.0, the value of the result at 5 different points is indicated by the shaded area below each point. Also, the vertical symmetry of f is the reason $f * g$ and $f \star g$ are identical in this example (From Wikipedia; author: Cmglee.).

## 6.2.7   Example

**Video-Analysis of Eye-Movement Recordings**

The human eye can rotate not only left/right and up/down, but also around the line-of-sight (dashed arrows in Fig. 6.10, left). While the horizontal/vertical eye movement can be determined quite easily from the position of the pupil, the measurement of this "ocular torsion" is more difficult. One way to determine it is by measuring the iris-pattern along a circular arc around the pupil center (solid gray line in Fig. 6.10, left) (Haslwanter and Moore 1995). Figure 6.10 shows the iral intensity when looking straight ahead, and in an eccentric eye position, when the sampling location is not adjusted for the 3D geometry of the eye. A *cross-correlation* can be used to determine the exact amount of shift of the whole pattern (from the location of the maximum), as well as information about how well the two patterns match (from the magnitude of the maximum).

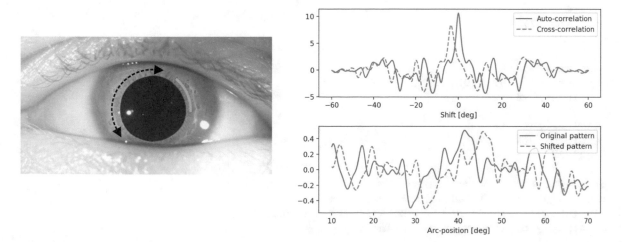

Figure 6.10: Simulated iris pattern, when looking straight ahead, and in an eccentric eye position. The pattern is similar, but shifted by approximately 3.5°.

## 6.3   Interpolation

When finding the point at which a curve crosses a given threshold we may want a higher accuracy than we have in the data. To do so we can find data points between recorded samples through *interpolation*.

What we discuss here is a simple approach, where we interpolate between fixed data points. In digital signal processing, interpolation is typically approached as a combination of low-pass filtering, followed by *decimation*. (Decimation by a factor of $n$ is if we take every $n$th data point from a signal.) This approach allows the specification of exact frequency responses. You should look these topics up for example if you down-sample an audio signal from a CD, where you want good control over the way your data manipulation affects the frequency content of the original signal (e.g. Smith 2010).

A typical application of interpolation is the combined analysis of two signals acquired at different sampling rates. In order to bring the two data sets to the same time-base, both can be interpolated with the same frequency. This also allows to incorporate signals sampled at irregular intervals, for example if data points were lost due to a bad wireless connection.

### 6.3.1   Linear Interpolation

The simplest form of interpolation is *linear interpolation*, where points between samples are obtained through linear connections between adjacent samples (Fig. 6.11, blue dashed line). While this is a computationally quick approach, it has two disadvantages:

- It is not very accurate.

- The first derivative of the resulting curve is discontinuous at the location of the samples.

Listing 6.3: interpolation.py

```
""" Linear and Cubic interpolations """

# Import the standard packages
import numpy as np
from scipy.interpolate import CubicSpline
import matplotlib.pyplot as plt
from scipy import signal

# Generate the data
x = np.arange(7)
y = np.sin(x)

# Linear interpolation
xi = np.arange(0, 6, 0.01)
yi = np.interp(xi, x, y)

# Cubic interpolation
cs = CubicSpline(x, y)
yic = cs(xi)

# Plot polynomial interpolations
plt.plot(x, y, 'ro', label = 'original data')
plt.plot(xi, yi, ls='dashed', label='linear interpolation')
plt.plot(xi, yic, label='cubic spline')

# Format the plot
ax = plt.gca()
ax.set_yticks(np.linspace(-1, 1, 5))
ax.axhline(0, LineStyle='dotted')
```

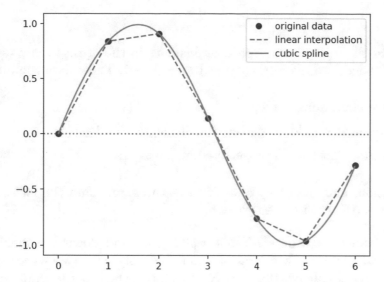

Figure 6.11: Cubic spline interpolation (from Listing 6.3).

```
plt.legend()

# Save and show the figure
out_file = 'interpolations.jpg'
plt.savefig(out_file, dpi=200, quality=90)
print(f'Image saved to {out_file}')

plt.show()
```

## 6.3.2   Cubic Spline Interpolation

The disadvantages of linear interpolation can be overcome with *cubic spline interpolation* (Fig. 6.11, orange solid line). Thereby the data points between samples are derived from cubic polynomials. The expression *spline* indicates that the polynomial coefficients satisfy two conditions:

- They are continuous at the location of the samples.

- The derivatives at the end of each polynomial are continuous up to a certain order.

With `sp.interpolate.CubicSpline`, the polynomial is chosen such that the first and the second derivative are continuous.

**Note:** *interpolation* splines are distinctly different from *B-splines*: While the former are such that they always go *through* the given data points, the latter are such that they are *attracted* by the data (see Sect. 5.5.2).

## 6.4    Exercises

1. **Event Finding** The file S6_1_data.npz contains data in the .npz-format (see Section 3.4), which can be loaded into the workspace with np.load. These data contain the keys

   **signal** *numpy* vector containing the signal,

   **step** the first feature that should be detected in the signal, and

   **sine** the second feature that should be detected in the signal.

   Use a cross correlation, and the step and sine features, and find the locations where each of these patterns occurs in the signal.

2. **Synchronization (more advanced)** While walking up and down one flight of spiral stairs, I have recorded the linear acceleration simultaneously with two sensors held close to my body: one my mobile phone (mobile_phone.txt), the other a commercial sensor (ngimu.txt). To synchronize the two sensors I have slapped the hand holding both sensors.
   Try to synchronize the two acceleration measurements to the point of maximum total acceleration (= slapping the hand). Then try to resample the acceleration data of both signals at 100 Hz, and store the result.

3. **EMG-Analysis**

   EMG-data are some of the most common signals in movement analysis. But sometimes the data analysis is not that simple. For example, data can be superposed by spurious drifts. And short drops in EMG activity can obscure extended periods of muscle contractions. The data in Shimmer3_EMG_Calibrated.csv have been taken from https://www.shimmersensing.com/support/sample-data/, where also the data description is given in detail.
   Write a function that does the following:

   - Import the EMG data from the data file Shimmer3_EMG_Calibrated.csv.
   - Remove the offset of the EMG-recording.
   - Rectify the data, and smooth them to produce a rough envelope of the signal.
   - Find the start- and end-points of muscle activity.
   - Eliminate artefacts.
   - Calculate and display the mean contraction time.

4. **Heart Rate Variability** Heart rate variability (HRV) is the physiological phenomenon of variation in the time interval between heartbeats. It is measured by the variation in the beat-to-beat interval. In clinical environments, it is used for the outcome prediction of a range of pathologies.
   One of the standard formats to store electro-cardiogram (ECG) data is the hea format, which can be read in Python with the package *wfdb*. Use this package to read in rec_1.dat, and calculate the HRV, defined here as the standard deviation of the time-differences between 10 RR-intervals, where R is a point corresponding to the peak of the QRS complex of the ECG wave. (The QRS complex is the combination of three of the graphical deflections seen on a typical ECG.[4]) Display the minimum and maximum HRV in this data set on the screen.

---

[4]For details see https://en.wikipedia.org/wiki/Electrocardiography.

# References

Haslwanter, T., & Moore, S.T. (1995). A theoretical analysis of three-dimensional eye position measurement using polar cross-correlation. *IEEE Transactions on Biomedical Engineering*, *42*(11), 1053–1061. https://doi.org/10.1109/10.250591.

Smith, III, J. O. (2010). *Physical audio signal processing: For virtual musical instruments and digital audio effects*. Stanford: W3K Publishing.

# Chapter 7

# Statistics

This chapter gives a very short introduction to the principles of statistical data analysis. The interpretation of confidence intervals for parameters, which will also be important in Chap. 8, is described through an explanation of the "standard error of the mean". In addition, common tests on the means of normally distributed signals, the so-called "T-tests", are introduced.

## 7.1 Statistical Basics

Measurement signals are inherently noisy. So when dealing with a given set of data, we always have to answer the questions "What do we know, based on the given set of data? And how certain are we of that knowledge?" Statistics provides us with the tools to extract the maximum amount of knowledge from given data. This chapter presents the basic principles of statistical data analysis.

The first step in the data analysis should always be a visual inspection of the data. It may be surprising to learn that approximately half the human cortex is dedicated to the analysis of visual information. As a result, we are tremendously good at seeing patterns in data, even though they may be hard to proof mathematically.

For the examples in this chapter, we take simulated weights and body-mass-indices (BMIs) of average male adults from Austria, China, and the USA.[1] The BMI is a simple index commonly used to classify overweight and obesity. It is calculated as a person's weight (in kg) divided by the height (in m) squared: for example, a person with a height of 1.75 m and a weight of 70 kg has a $BMI = \frac{70}{1.75^2} = 22.9$. A $BMI > 25$ indicates overweight. Although it is only a crude measure of fatness, it gives a good reference value for the average adult in a country.

Based on the simulated measurement of the BMI from 50 adult male subjects of a given country, what can we say about the weight of the male adult population in that country, and how does it compare to other countries? This chapter will show how to answer such questions.

### 7.1.1 Principles of Inferential Statistics

The basic idea of *inferential statistics* is presented in Fig. 7.1: We want to know something about a parameter in a whole population. For example, we want to know the BMI of the average Austrian male adult. (The parameter of interest, in this example the BMI, is sometimes called *statistic*.) But we cannot measure all Austrians. The best thing we can do is measure a representative *sample* of them. Based on the values in the sample (the mean sample BMI, the variability, the distribution, etc.) we then use *inference* to find out what we can say about the corresponding parameter in the population, and how certain we are about that knowledge.

---

[1] Data taken from WHO 2016 https://www.who.int/data/gho/data/indicators/indicator-details/GHO/mean-bmi-(kg-m-)-(age-standardized-estimate).

T. Haslwanter, *An Introduction to Hands-on Signal Analysis with Python*,
https://doi.org/10.1007/978-3-030-57903-6_7

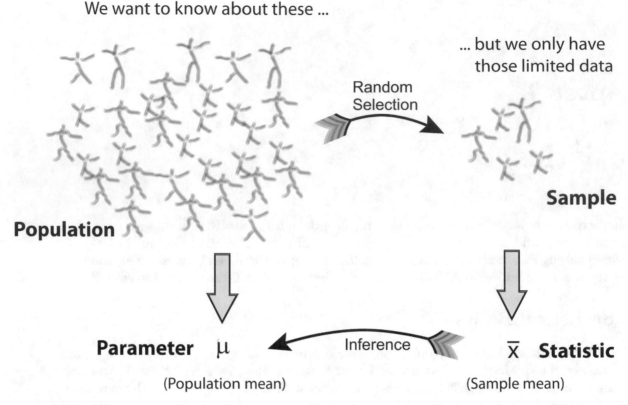

Figure 7.1: With *Statistical Inference*, information from a *Sample* is used to estimate *Parameters* from the corresponding *Population*. (from Haslwanter (2016), with kind permission from Springer Publishing).

## 7.1.2   Common Statistical Parameters

### Mean and Range

In scientific publications data are usually described by their mean and standard deviation, with the sample range often in brackets. For example, *"The age of our subjects was* $74 \pm 8$ *years (range 66–85 years)."*

For symmetrical distributions such as a bell-curve, the *sample mean* is the best estimate for the typical population value the we can obtain from our data. It is commonly indicated with a bar above the variable name

$$\bar{x} = \frac{\sum\limits_{i=1}^{n} x_i}{n} \tag{7.1}$$

The *range* is the upper and lower limit of the data. It is often also quoted in publications, in addition to mean and standard deviation, so that the reader can get a quick numerical overview of the data values.

### Median, Quartiles, and Centiles

For asymmetrical distributions, additional information can be obtained through sorting the data by value, and splitting them in four parts:

**Upper quartile** Smallest value that is larger than 75% of all data.

**Median** Smallest value that is larger than 50% of all data.

**Lower quartile** Smallest value that is larger than 25% of all data.

If enough data are available, the sorted data can be split into 100 parts. This gives the so-called "centiles" or "percentiles". For example, the $5^{th}$ *percentile* is the smallest value that is larger than 5% of all data. In Python, centiles can be found with

```
centile = np.percentile(data, percent_value)
```

When a data distribution is symmetrical, the mean and the median of the data approximately coincide. But for asymmetrical distributions there can be a significant difference. In that case the median is more representative of the typical value, as it is robust against individual outliers. As a result, a running median is a powerful way to eliminate the effect of outliers (see Fig. 5.9). To illustrate the difference between mean and median, take for example the mean income of the visitors of a well frequented American bar: if Bill Gates happens to walk into that bar, the mean income of the visitors will go up dramatically. In contrast, the median income will stay approximately the same.

*Box plots* use quartiles to describe asymmetrical sample data, or sample data with distinct outliers (Fig. 7.2):

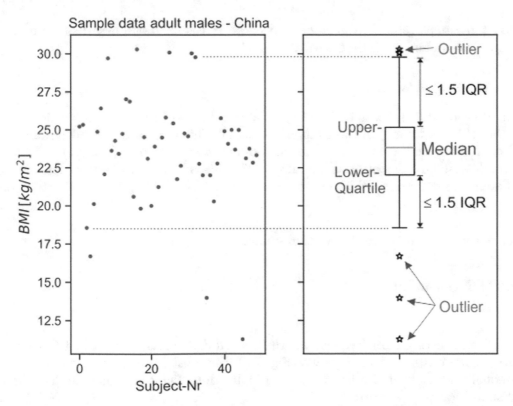

Figure 7.2: **Left:** Scatter plot **Right:** Corresponding box plot. The difference between the upper and lower quartile is called the *Inter-Quartile-Range (IQR)*. Error-bars commonly indicate the most extreme values inside *1.5\*IQR* outside the upper/lower quartile.

Care has to be taken with the whiskers, as different conventions exist for them. The most common form is that the lower whisker indicates the lowest value within 1.5 * *inter-quartile-range* (IQR) of the lower quartile, and the upper whisker the highest value within 1.5 * IQR of the upper quartile.[2] As suggested by Tukey, data samples outside these whiskers are defined as "outliers" and are plotted separately (Tukey 1977).

---

[2] Another convention sometimes used is to have the whiskers indicate the full data range.

**Outliers**

The correctness of outliers should always be carefully checked, as they might contain erroneous values. For experimental studies this can be checked by looking up the corresponding entries in the experimental log files, to see if experimental anomalies could explain these outliers. In that case the corresponding data points may be eliminated. In all other cases they have to be kept(!), and often provide important information about the underlying population.

**Standard Deviation and Variance**

The *standard deviation (SD)* takes us back to the population: we have to carefully distinguish between the "standard deviation of our sample" and the "standard deviation of the population", because the two in general do *not* coincide (see Fig. 7.1)! It is common practice to use Greek letters for the population means and standard deviations, $\mu$ and $\sigma$, but Latin letters for the corresponding sample parameters ($m$ or $\bar{x}$, and $s$.)

Based on our data, the best estimate we can get for the expected standard deviation of the *population*(!) is

$$s = \sqrt{\frac{\sum (w_i - \bar{w})^2}{n - 1}} \tag{7.2}$$

This is called the *sample standard deviation* (because it is our best guess based on the sampled data), and is obtained in Python with

```
sd = np.std(data, ddof=1)
```

Note that the division in Eq. 7.2 is by `(n-1)`, which is reflected in the parameter `ddof=1` ("difference in degrees of freedom") in the Python command. This is not quite intuitive, and is caused by the fact that the real mean $\mu$ is unknown, and our *sample mean* $\bar{x}$ (Eq. 7.1) is chosen such that the sample standard deviation is minimized. This leads to an underestimation of the true population standard deviation, which can be compensated by dividing by $n - 1$ instead of $n$.

The *variance* is simply the standard deviation squared.

### 7.1.3 Normal Distribution

**Definition**

The data distribution occurring most frequently is the *normal distribution*, also called *Gaussian distribution*, with its famous bell-shaped profile (Fig. 7.3): the most likely value is the average value, and it is equally likely to find a low value or a high value. The mathematical definition of the normal distribution is:

$$f_{\mu,\sigma}(x) = \frac{1}{\sigma\sqrt{2\pi}} e^{-(x-\mu)^2/2\sigma^2} , -\infty < x < \infty \tag{7.3}$$

$f_{\mu,\sigma}$ is called the *probability density function (PDF)* of the normal distribution. As indicated in Fig. 7.4, the integral of the PDF between $a$ and $b$ is the probability that the next sample value will lie in that interval.

For a normal distribution the *mean value* $\mu$ indicates the location of the distribution, and the *standard deviation* $\sigma$ its variability or "scale". Mean and standard deviation completely characterize the normal distribution.

Generating a normal distribution and plotting the corresponding PDF is straightforward:

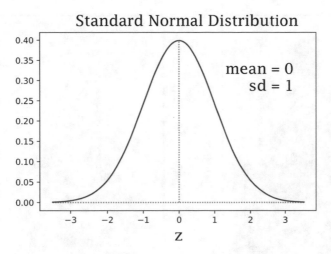

Figure 7.3: A normal distribution with a mean of 0 and a standard deviation of 1 is called a *standard normal distribution* or *z-distribution*.

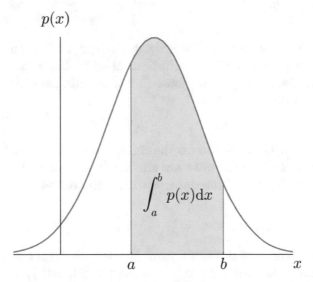

Figure 7.4: Interpretation of the *Probability Density Function (PDF)*: the integral of the PDF between $a$ and $b$ is the probability that the next sample value will lie in that interval.

```python
# Generate the probability distribution, here e.g. for the weight
# of adult Austrian men
from scipy import stats

mu = 75
sd = 12

# In Python a distribution with all parameters fixed is called
# a "frozen distribution function"
nd = stats.norm(mu, sd)

# Calculate and plot the corresponding PDF-function
x = np.arange(mu-3*sd, mu+3*sd)
pdf = nd.pdf(x)
plt.plot(x, pdf)
```

The PDF describing a normal distribution is closely related to the histogram of data samples. The *frequency histogram* (Fig. 7.5, left) indicates the counts in each bin. Dividing by the total

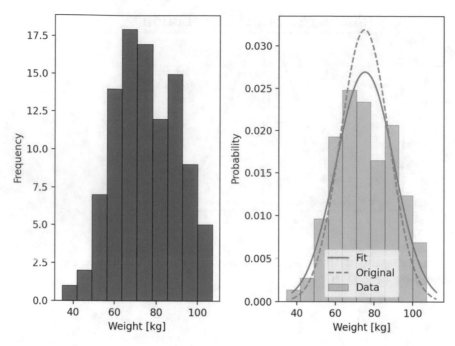

Figure 7.5: **Left:** Frequency histogram of simulated body weight of 100 subjects, assuming those weights are noisy values from a normal distribution. **Right:** Corresponding probability histogram. The solid line shows the corresponding best-fit normal distribution, and the dashed line indicates the original normal distribution.

number of counts gives the *probability histogram* on the right. Note that by definition, the area in a probability histogram is exactly *1*. If the data are similar to a normal distribution, we can calculate the best-fit normal distribution to the data, with the command

```
mean_fit, std_fit = std.norm.fit(data)
```

Here the fit is pretty good, and the data can now be well characterized by only a few parameters—here the mean and standard deviation of the normal distribution. As a result, such distributions are called "parametric data distributions".

## 7.2   Confidence Intervals

### 7.2.1   For Data

The *x-% confidence interval (CI)* of a distribution indicates the data range around the mean that contains x-% of all data. For a normal distribution the two-tailed confidence interval is given by (Fig. 7.7):

$$CI_{data} = \mu \pm \sigma * z_{1-\alpha/2} \tag{7.4}$$

where $z_{1-\alpha/2}$ indicates the percentile function of a standard normal distribution ("z-distribution"), which takes a %-value as input, and returns the corresponding percentile of the distribution (see Section 7.1.2). (For details see e.g. Haslwanter (2016).)

For a normal distribution, $\pm 1\,SD$ around the mean contains *2/3* of the data; $\pm 2\,SD$ contain approximately *95%*; and $\pm 3\,SD$ contain approximately *99.9%* (Fig. 7.6).

For the exact 95%-CI we set $\alpha = 0.05$ in Eq. 7.4. So if we want to know the 95%-CI, we have to calculate the corresponding z-values for $\alpha/2$ and for $1 - \alpha/2$: the factor $\alpha/2$ is due to the fact that we have outliers below and above the limits of the confidence intervals, and is called

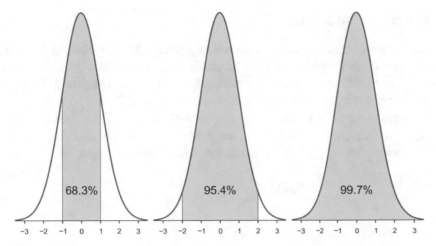

Figure 7.6: Normal Distribution: the unit on the horizontal axis is 1 standard deviation (SD). Data points lying outside +2 SD are called "significantly different"; data points outside ±3 SD are called "highly significantly different".

"two-sided" or "two-tailed confidence interval" (Fig. 7.7). In Python we can either calculate it directly:

```
CI = nd.interval(1-alpha)
```

or, equivalently, separately for the upper- and lower-limit with `.ppf` (*percentile point function*)

```
CI_lower = nd.ppf(alpha/2)
CI_upper = nd.ppf(1 - alpha/2)
```

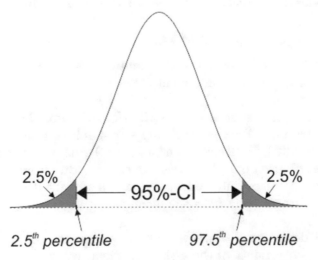

Figure 7.7: If 95% of the data are within the confidence limits, 2.5% of the data lie above, and 2.5% below those limits.

## 7.2.2 Standard Error of the Mean

The most important value characterizing an approximately normally distributed population is the mean value. This section describes how we can quantify the uncertainty in that parameter, and how this can be used for statistical tests. We will see in the next chapter (Chap. 8) that the same principle can be use for general tests of fitted parameters (Fig. 7.8).

The more subjects we measure, the more accurate our estimation of the mean value of our population will be. The remaining uncertainty is often referred to as the *standard error of the mean (SEM)* or short *standard error*, and is for Gaussian distributions given by

$$SEM = \frac{\sigma}{\sqrt{n}} \tag{7.5}$$

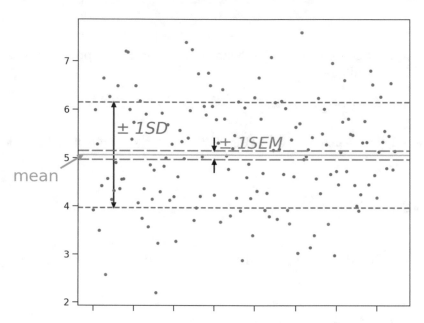

Figure 7.8: The *standard deviation (SD)* (blue short-dashed lines) describes the variability of the data. The *standard error of the mean (SEM)* (orange long-dashed lines) describes the uncertainty of the mean value.

The equation for the *confidence intervals for the mean value* is given by

$$CI_{mean} = \mu \pm SEM * t_{n-1,1-\alpha/2} \tag{7.6}$$

where $t_{n-1,1-\alpha/2}$ is the so-called "T-distribution for n samples", at an $\alpha$-% confidence level. This is very similar to Eq. 7.4, but with two important changes:

1. While the normal distribution describes the distribution of data, the mean value over $n$ samples from that distribution does—surprisingly—not follow a normal distribution. Instead it follows the so-called "T-distribution" (Fig. 7.9). The T-distribution is very similar to the normal distribution but also takes the errors induced by the estimation of the mean value based on $n$ subjects into consideration. As a result it has "larger wings", i.e. slightly higher values further away from the mean.[3]

2. And while the standard deviation $\sigma$ characterizes the variability of data, the standard error $SEM$ describes the uncertainty of the estimated mean value.

---

[3]In the rare case that the true value of $\sigma$ is known for the normal-distribution, the distribution of the mean is also described by a normal distribution.

The easiest way to calculate the standard error is

```
from scipy import stats
sem = stats.sem(data)
```

The standard T-distribution (for a mean of 0, and a standard error of 1) requires one argument, the *degrees of freedom (DOF)*. For the average over $n$ samples the DOF is given by `dof = n-1`, because one parameter (the mean) is already determined.

The required t-values for Eq. 7.6 can be obtained with `ppf` (for *percentile point function*). For example, for the case of five subjects and $\alpha = 0.05$ it is given by:

```
td = stats.t(df=4)
alpha = 5/100
tval =  td.ppf((1-alpha)/2)                 # Result: tval = 2.78
# or in one line, now for 21 subjects
tval =  stats.t(df=20).ppf((1-alpha)/2)  # Result: tval = 2.09
zval = stats.norm().ppf((1-alpha)/2)      # Result: zval = 1.9600
```

For five subjects we get a value of 2.78, and for 21 subjects a value of 2.09, which is already pretty close to the corresponding value from the normal distribution, 1.96.

Since this interpretation of confidence intervals is quite important, I want to spell it out explicitly once more:

> The *standard deviation* describes the variability of the data; and the corresponding 95% confidence interval (CI) for the data is the area around the mean that contains 95% of all data points. In contrast, the *standard error of the mean* describes our uncertainty about the mean value; and the corresponding 95% CI is the area that contains the true mean value with a probability of 95%.
>
> If a value lies outside the 95%-CI, we say that our value is "significantly different" from $x$. If it lies outside the 99.9%-CI, we call it "highly significantly different".

## 7.3 Comparison Tests for Normally Distributed Data

In the following we restrict our discussion to the analysis of data that are normally distributed. Such assumptions also have to be checked—at least visually! If these assumptions don't hold you have to use statistical tests that don't rely on these assumptions. For a more general introduction to statistical data analysis see for example Haslwanter (2016).

When evaluating the mean value of one or two groups, three cases can be distinguished:

1. Comparison of one group to a fixed value (Fig. 7.10)

2. Comparison between two independent groups (Fig. 7.12).

3. Comparison between two related groups (e.g. before-after experiments, Fig. 7.13).

### 7.3.1 Comparing Data to a Fixed Value

To continue with our example with the BMI, one question that could be asked is "Can Chinese men eat more or should they eat less?" To answer it, we compare the measured BMIs with the value *25*, which is the threshold between normal and over-weight. A visual inspection of our—simulated—data (Fig. 7.10) shows that while some Chinese do have a BMI above 25, most of them are below.

The statistical confirmation is obtained with

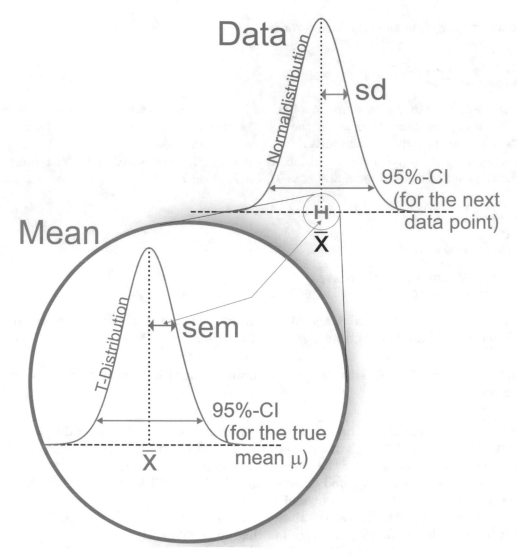

Figure 7.9: While the normal distribution describes the variability of the data, the T-distribution describes the variability of the mean value.

```
t, p = stats.ttest_1samp(data, ref_value)
```

where $p = 0.008$. To interpret the return arguments of `ttest` we first have to look at the concept of "hypothesis tests".

### 7.3.2   Hypothesis Tests

We could rephrase the question from above "Can Chinese men eat more or should they eat less?" (Fig. 7.10) with "Is the measured Chinese BMI different from the threshold value of *25*?" But what would that mean, exactly? We would obviously regard a BMI of *30* as clearly more than *25*. But would we also regard a BMI of *25.1* as more? And what about *25.00001*?

To get around that quandary hypothesis tests turn the logic around. They always start with a *null hypothesis*, i.e. assuming that there is *null* difference between the value of interest and the value in the tested group. If the statistical analysis indicates that there is only a very small probability to get a data set such as the one tested under this null hypothesis, then we say that the tested data "differ significantly" from the value of interest.

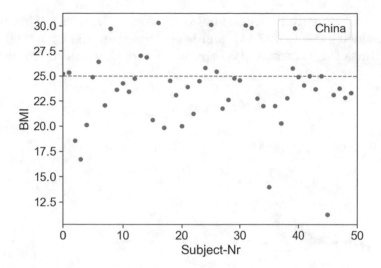

Figure 7.10: Data can be compared to a value with `ttest_1samp` (from `F7_stats_examples.py`).

This explains the return values (`t`, `p`) from the command `ttest_1samp` in the example above: `t` corresponds to the "t-statistic" of the data set, i.e. the difference from the true mean scaled by the standard error of the mean (SEM). And

> `p` is the probability to obtain a value as extreme or more extreme than the observed one, if the null hypothesis is true.

This p-value is typically quoted in the Results section of scientific articles.

In the example above, the null hypothesis would be: "We hypothesize that *(average BMI—25)* is equal to null." Since p is very small here (0.008), this null hypothesis has to be rejected. By convention, the following thresholds are commonly used for the interpretation of p:

- If $p < 0.05$ the null-hypothesis is rejected, and we speak of a *significant difference*.

- If $p < 0.001$ the null-hypothesis is rejected, and we speak of a *highly significant difference*.

- Otherwise the null-hypothesis is accepted, and we state that *there is no significant difference*.

**Tip:** Instead of reporting only *p-values*, it has become recommended practice to also report the confidence intervals of tested parameters. If the null hypothesis is that the measured value is zero, the hypothesis has to be accepted if the 95%-confidence interval includes zero. Otherwise, the hypothesis has to be rejected. (This is important for parameter fitting, see Sect. 8.5.2.)

### 7.3.3 One-sided versus Two-sided Comparisons

By far most statistical problems require a *two-sided comparison*, and correspondingly the "two-tailed" or "two-sided confidence intervals" (as indicated in Fig. 7.7) are appropriate. However, if the data set in one group can only be larger (or only be smaller) than the data set from the other group, a *one-sided comparison* has to be used.

Take for example the case that you have bought an expensive Swiss chocolate cookie, and suspect that it contains "too little" chocolate. In that example you you don't want to check if it is "different" from normal chocolate cookies (i.e. too much or too little chocolate, which would be indicated by the two-tailed confidence interval). But you want to check if it is in the lowest

5% of the chocolate probability distribution (one-tailed confidence interval). This corresponds to a "one-tailed" or "one-sided" test (Fig. 7.11), and is controlled by the command parameter `alternative`. Valid settings for `alternative` are: `two-sided`, `less`, or `greater`.

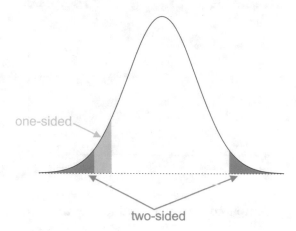

Figure 7.11: If we do NOT know in advance if we are looking for samples that are larger or smaller than the typical sample, we have to use *two-sided t-tests*. But if we know in advance that only deviations in one direction is relevant, then a *one-sided comparison* should be used.

### 7.3.4  Comparing Two Independent Groups

The second case we want to discuss here is the comparison of two independent groups. For example, we take measured BMIs from 50 Austrians and 50 Germans and ask the question "Do Germans have a different BMI than Austrians?"

Two independent groups are compared with the *t-test for independent samples*, `ttest_ind`:

```
t, p = stats.ttest_ind(Austrians, Germans)
```

For the data shown in Fig. 7.12 the mean BMI for the German subjects (dashed line) is larger than the mean value for the Austrians (dash-dotted line). For these data `ttest_ind` gives a p-value of `0.67`, and our conclusion has to be "Based on our data, there is no significant difference between Austrians and Germans." The variability in the data is so big that the measured difference could have arisen by chance alone.

If we want to detect small differences between two independent groups, we need to test a larger number of subjects since the standard error becomes smaller with increasing sample size (Eq. 7.5).

### 7.3.5  Pre-Post Comparisons

The analysis has to be different when comparing two groups where every data point in the first group directly corresponds to a specific data point in the second group. For example, assume all subjects in our German group go on a diet for one week, and are then weighed again. Further assume that the diet is working pretty well, and most subjects lose some weight—but only a very small amount. If we would use `ttest_ind`, that small change would be completely swamped by the variability between the subjects. But if we take the before-after difference we will see a consistent, significant weight reduction. This is a so-called *paired t-test*, and is automatically performed when the command `ttest_rel` is used with two groups of the same size. So the following two following commands give the same results (Fig. 7.13):

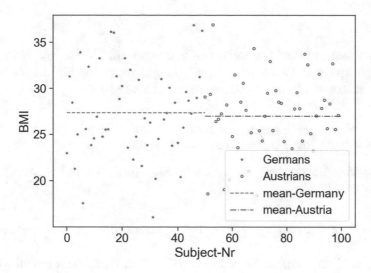

Figure 7.12: Comparison of two *independent* groups with `ttest_ind` (from `F7_stats_examples.py`).

```
t, p = stats.ttest_rel(preDiet, postDiet)
t, p = stats.ttest_1samp(preDiet-postDiet, 0)
```

In other words, comparing two related groups gives the same results as comparing the paired difference between them to zero. Due to its nature `ttest_rel` can detect smaller significant differences than `ttest_ind`.

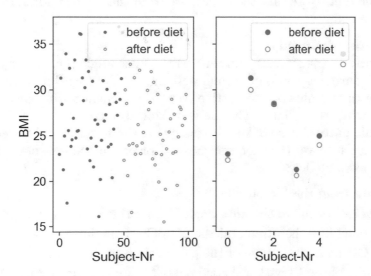

Figure 7.13: `ttest_rel` is used for *paired T-tests*, e.g. pre/post comparisons (from `F7_stats_examples.py`).

## 7.4   Exercises

1. **Analyze data** Let us assume that the weight of 200 random, 15 year old children, sorted randomly into two groups are: (To ensure that you get reproducible numbers, initialize the random-seed-generator with `np.random.seed(12345)`)

```
group_1  = 7 * np.random.randn(100) + 60
group_2 = 10 * np.random.randn(100) + 55
```

Then the children in group2 are put on a "carb-diet" (only carbohydrates). After one month these children weigh

```
group1_after = group1 - 1.5 + 1.0 * np.random.randn(len(group1))
```

   • calculate mean, median, standard deviation (SD), and standard error for the first 10 children in group1.
   • calculate mean, median, standard deviation (SD), and standard error for all children in group1.
   • How do you interpret the standard errors here?

2. **Errorbar plots**

   • Plot mean +/- SD , and mean +/- SE for the first 10 children in group1.
   • Plot mean +/- SD , and mean +/- SE for all children in group1.

3. **Box plots**

   • Make box plots of the children in group1 and in group2, and display them on the same axis.

4. **Compare groups**

   • Find out if the difference between the children in group1 and the children in group2 is significant.
   • Find out if the weight of the children in group1 after the carb-diet is less than the weight before the diet. Note that these data are from the same children as before the diet, in the same order.

5. **Gait Analysis (more advanced)**
   The recording in `.\data \gait.mat` contains (i) the knee angle (i.e. the angle between upper- and lower leg) during a period of normal walking, (ii) the recording time, and (iii) a vector containing the time-points of "heel-strike", i.e. the moment during each 'gait cycle' that the heel first touches the ground. The file is in Matlab format.
   Write a script that calculates the mean knee angle during a gait cycle, and the corresponding the 95%-confidence interval (CI) for the mean. Normalize the gait-cycle to "100%" (see Fig. 7.14). Proceed as follows

   • Reads in the data from the Matlab file.
   • For each gait cycle, normalize the knee-angle during this cycle, to 101 points.
   • For each point, calculate the mean and the 95% CI of the knee angle.
   • Plot mean and CIs as a percentage of the gait cycle (see Fig. 7.14).
   • Can you also indicate the CI with a shaded patch? (Tip: use the *matplotlib* command `plt.fill_between`.)

For the confidence interval of the mean, first calculate the standard error of the mean (SEM), and then use the fact that the 95% CI is approximately $\pm 2\,SEM$.

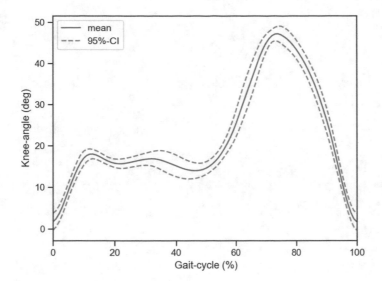

Figure 7.14: 95% confidence-interval for the knee angle, during a normalized gait-cycle.

# References

Haslwanter, T. (2016). *Introduction to Statistics with Python.* Berlin: Springer.

Tukey, J. W. (1977). *Exploratory Data Analysis.* Reading: Addison-Wesley.

# Chapter 8

# Parameter Fitting

Many data sets can be well approximated by a simple model. If this is the case, it tremendously simplifies the description of the data: we do not need to know every single data point any more, but only the coefficients defining the best-fit model to the data points. This chapter describes how to obtain the model coefficients for simple linear models, how accurately we know these coefficients, and how this concept can be used to describe much more complex patterns, such as polynomials, circles, etc.

## 8.1 Correlations

A *correlation* between two parameters describes any statistical relationship between them. It answers the question: "When I change one variable, which percentage of the change of the other variable is determined, and which percentage is unknown?"

### 8.1.1 Correlation Coefficient

The *correlation coefficient* $r$ is a measure of the linear correlation (or dependence) between two variables $x$ and $y$, giving a value between $+1$ and $-1$. A value of $+1$ indicates a perfect positive correlation, $0$ is no correlation, and $-1$ a perfect negative correlation. Figure 8.1 shows examples of data sets with different correlation coefficients (Fig. 8.1). For sample data $x_i$ and $y_i$, $r$ is given by the change in $x$ multiplied by the change in $y$, normalized by the relative spread of $x$ and $y$:

$$r = \sum_{i=0}^{n-1} \left( \frac{x_i - \bar{x}}{\sqrt{\sum_{j=0}^{n-1}(x_j - \bar{x})^2}} * \frac{y_i - \bar{y}}{\sqrt{\sum_{k=1}^{n-1}(y_k - \bar{y})^2}} \right) \tag{8.1}$$

Note that—in contrast to linear regression fits—the correlation coefficient is symmetric in $x$ and $y$.

© The Author(s), under exclusive license to Springer Nature Switzerland AG 2021
T. Haslwanter, *An Introduction to Hands-on Signal Analysis with Python*,
https://doi.org/10.1007/978-3-030-57903-6_8

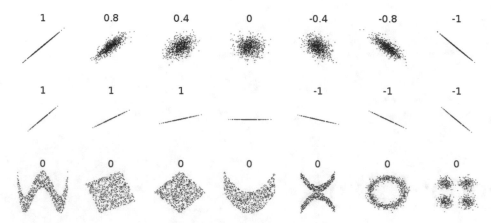

Figure 8.1: The correlation coefficient only quantifies how well points lie on a *straight line*, and if that line is *rising or falling*. Other relationships are not captured (see bottom row) (Illustration taken from DenisBoigelot, Wikipedia).

### 8.1.2   Coefficient of Determination

For linear regression, the square of the correlation coefficient, $r^2$, is called the *coefficient of determination*, and quantifies how well the fitted data account for the raw data. It makes use of the fact that

$$SS_{tot} = SS_{model} + SS_{resid} \tag{8.2}$$

where $SS_{tot}$ is the *sum of the squares of the total deviation* (sum of red boxes, Fig. 8.2, left), $SS_{model}$ the *sum of the squares of the model fit*, and $SS_{resid}$ the *sum of the squares of the residuals* (sum of blue boxes, Fig. 8.2, right). These values are given by:

$$SS_{tot} = \sum_i (y_i - \bar{y})^2 \tag{8.3}$$

$$SS_{model} = \sum_i (\hat{y}_i - \bar{y})^2 \tag{8.4}$$

$$SS_{resid} = \sum_i (y_i - \hat{y}_i)^2 \tag{8.5}$$

where $\bar{y}$ is the average value, and $\hat{y}_i$ is the value of the fitted data point that corresponds to a given $x_i$ value.

The coefficient of determination is defined as the amount of the sum of squares that can be explained by the model:

$$r^2 = \frac{SS_{model}}{SS_{tot}} = 1 - \frac{SS_{resid}}{SS_{tot}}$$

As an example, $r^2 = 0.7$ would indicate that *approximately seventy percent of the variation in the response can be explained by the line fit.*

To illustrate a range of different fits, three data sets with different amounts of noise are plotted in Fig. 8.3. All three data sets have the same line of best fit, but the coefficients of determination differ.

The correlation coefficient r, sometimes also called *Pearson's correlation coefficient*, can be calculated with `scipy.stats.pearsonr`. More complete information about the linear regression can be obtained with `scipy.stats.linregress`. The code below produces data with a correlation coefficient of approximately $r = 0.997$.

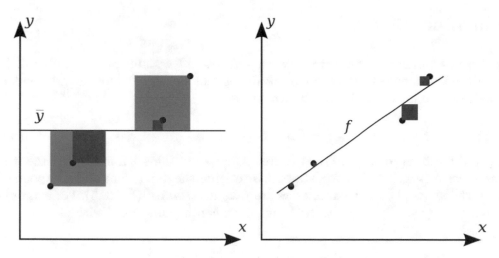

Figure 8.2: The better the linear regression (on the right) fits the data in comparison to the simple average (on the left), the closer the value of $r^2$ is to 1. The areas of the blue squares (right) represent the squared residuals with respect to the linear regression, $SS_{resid}$. The areas of the red squares (left) represent the squared residuals with respect to the average value, $SS_{tot}$ (from Wikipedia).

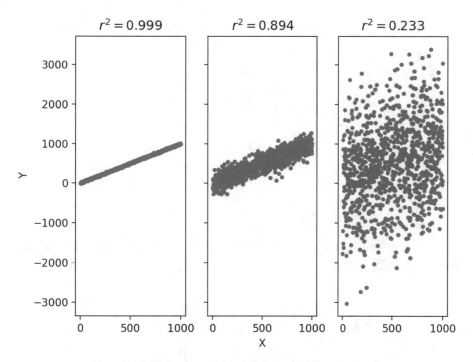

Figure 8.3: Data with different levels of noise.

```
import numpy as np
import matplotlib.pyplot as plt
from scipy import stats

x = np.arange(100)
y = 10 + 0.5*x + np.random.randn(len(x))
plt.plot(x,y,'.');

slope, intercept, r_value, p_value, std_err = stats.linregress(x, y)
```

## 8.2   Straight Lines

A linear model describes a linear relationship between a *dependent* variable and a number of *independent* variables. In the simplest case, there is one dependent variable $y$ and one independent variable $x$. In that case the linear model can be written as

$$y = mx + b \tag{8.6}$$

where $m$ ("multiplier") and $b$ ("bias") are the two free parameters. This is a line in a plane with *slope $m$* and *$y$-intercept $b$* (Fig. 8.4). If two data points relating the dependent and independent variables are given, then these two free parameters are *fully determined* (Fig. 8.5). For example, if $P_1 = (x_1/y_1)$ and $P_2 = (x_2/y_2)$, then intercept and slope can be calculated with

$$m = \frac{\Delta y}{\Delta x} = \frac{y_2 - y_1}{x_2 - x_1}, \text{ and} \tag{8.7}$$

$$b = \frac{y_1 x_2 - x_1 y_2}{x_2 - x_1} \tag{8.8}$$

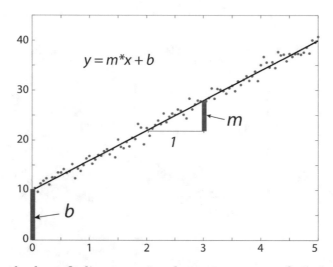

Figure 8.4: To obtain the best-fit line to noisy data, we want to find the model coefficients $m$ and $b$ that define our data-model $y = m * x + b$.

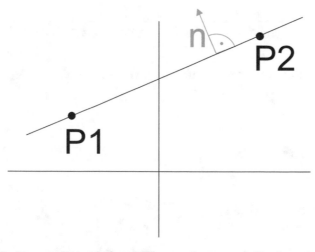

Figure 8.5: Two points define a line. **n** is perpendicular to that line.

In the general case there is one dependent variable $y$ and $p + 1$ independent variables, which are often indicated with $\beta_i$. The linear model then becomes

$$y = \beta_0 + \sum_{i=1}^{p} \beta_i x_i \tag{8.9}$$

where $\beta_0, \beta_1, ..., \beta_p$ are the $p + 1$ free parameters. If we have $p + 1$ different data points that lie exactly on a k-dimensional hyperplane, then the free parameters are fully determined.

### 8.2.1 Normal Form of Line Equation

A line can either be represented by its slope and y-intercept, as shown in Eq. 8.6. Alternatively, it can be characterized by a point on the line, $\mathbf{x_1}$, and a vector $\mathbf{n}$ that is perpendicular to the line (see Fig. 8.5):

$$\mathbf{n} \cdot \mathbf{x} = \mathbf{n} \cdot \mathbf{x}_1 \tag{8.10}$$

or

$$ax + by = c, \text{ with } \mathbf{n} = \begin{pmatrix} a \\ b \end{pmatrix} \text{ and } c = \mathbf{n} \cdot \mathbf{x_1} \tag{8.11}$$

The advantage of Eq. 8.10 is that it holds not only for a 1D line in a 2D plane, but also for a 2D plane in 3D space, etc.

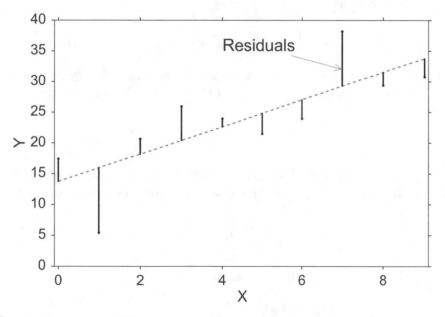

Figure 8.6: *Residuals* are the difference between data and modeled value, here $y = m * x + b$. Note that the values along the x-axis are fixed!.

## 8.3 Line Fitting

Typically we have more data points than free parameters. In that case the model is *over-determined* by the data, so that it is not possible to fit the model exactly to all data points. Instead we try to find a *best fit model*, which is the model with the minimum fitting error. Often the fitting error is taken to be the *least squares estimator* (see Sect. 8.3.2).

### 8.3.1   Residuals

The difference between a data value and the corresponding model value is called a *residual* (Fig. 8.6). If we have a data point where the dependent variable has the value $y_k$ and the corresponding independent variables have value $x_{ki}$, then the residual $e_k$ for the general linear model is

$$e_k = y_k - \left( \beta_0 + \sum_{i=1}^{p} \beta_i x_{ki} \right) \tag{8.12}$$

### 8.3.2   Least Squares Estimators

If there are $n$ data points, the sum of the squared residuals is

$$E(\mathbf{x}, \mathbf{y}, \boldsymbol{\beta}) = \sum_{k=0}^{n-1} e_k^2 = \sum_{k=0}^{n-1} \left( y_k - \left( \beta_0 + \sum_{i=1}^{p} \beta_i x_{ki} \right) \right)^2 \tag{8.13}$$

The *least squares estimators* are the values $\hat{\beta}_i$ that minimize $E$. To determine the value of the least squares estimators $\hat{\beta}_i$ it is necessary to locate the minimum of $E$ by finding where the following partial derivatives are zero:

$$\begin{aligned}
0 &= \left. \frac{\partial E}{\partial \beta_0} \right|_{\hat{\beta}} = -2 \sum_{k=0}^{n-1} \left( y_k - \left( \hat{\beta}_0 + \sum_{i=1}^{p} \hat{\beta}_i x_{ki} \right) \right) \\
0 &= \left. \frac{\partial E}{\partial \beta_i} \right|_{\hat{\beta}} = -2 \sum_{k=0}^{n-1} \left( y_k - \left( \hat{\beta}_0 + \sum_{i=1}^{p} \hat{\beta}_i x_{ki} \right) \right) * x_{ip} \quad \forall\, i = 1, .., p
\end{aligned} \tag{8.14}$$

This linear system of equations can be solved to find the values of the least squares estimators. Luckily we don't have to do that by hand, and it can be done with by the Python functions describe in the next section.

**Ordinary Least Squares**

The method of ordinary least squares can be used to find an approximate solution to overdetermined systems. For the system $\mathbf{A} \cdot \mathbf{p} = \mathbf{y}$, the least squares formula is obtained from the problem

$$\min_{p} \| \mathbf{A} \cdot \mathbf{p} - \mathbf{y} \|,$$

the solution of which can be written with the normal equations,

$$\mathbf{p} = (\mathbf{A}^{\mathrm{T}} \mathbf{A})^{-1} \mathbf{A}^{\mathrm{T}} \cdot \mathbf{y}, \tag{8.15}$$

where T indicates a matrix transpose, provided $(\mathbf{A}^{\mathrm{T}} \mathbf{A})^{-1}$ exists (that is, provided $\mathbf{A}$ has full column rank) (see e.g. https://en.wikipedia.org/wiki/Linear_regression). With this formula an approximate solution is found when no exact solution exists, and it gives an exact solution when one does exist.

## 8.4   Linear Fits with Python

### 8.4.1   Linear Model without Intercept

Let us start with the simplest case, where there is no intercept term (i.e., $\beta_0$ is zero). Then the model is

$$y \approx \sum_{i=1}^{p} \beta_i x_i \,. \tag{8.16}$$

With $\boldsymbol{X}$ an $n \times p$ matrix made of the column vectors $\mathbf{x_1}, ..., \mathbf{x_p}$, $\boldsymbol{\beta}$ is a $p \times 1$ column vector containing the free parameters $\beta_1, ..., \beta_p$, and $\mathbf{y}$ an $n \times 1$ column vector, we can rewrite the model as

$$\mathbf{y} \approx \boldsymbol{X} \cdot \boldsymbol{\beta}$$

The least squares estimator for $\boldsymbol{\beta}$ can be obtained algebraically by reshaping the equation:

$$\hat{\boldsymbol{\beta}} = \boldsymbol{X}^{-1} * \mathbf{y} \tag{8.17}$$

Since in general the exact inverse of $\boldsymbol{X}$ does not exist, we have to use the "pseudo-inverse"

```
m = np.linalg.pinv(X) @ y
```

or, to the same effect

```
m = np.linalg.lstsq(X, y)[0]
```

## 8.4.2   Linear Model with Intercept

The general linear model with intercept term is given by

$$y_k \approx \left( \beta_0 + \sum_{i=1}^{p} \beta_i x_{ki} \right)$$

or in matrix form

$$\mathbf{y} \approx \mathbf{X} \cdot \boldsymbol{\beta} \tag{8.18}$$

with the following matrix and vector definitions:

$$\mathbf{X} = \begin{bmatrix} 1 & x_{11} & x_{12} & \cdots & x_{1p} \\ 1 & x_{21} & x_{22} & \cdots & x_{2p} \\ \vdots & \vdots & \vdots & & \vdots \\ 1 & x_{n1} & x_{n2} & \cdots & x_{np} \end{bmatrix}, \boldsymbol{\beta} = \begin{bmatrix} \beta_0 \\ \beta_1 \\ \vdots \\ \beta_p \end{bmatrix}, \mathbf{y} = \begin{bmatrix} y_1 \\ y_2 \\ \vdots \\ y_n \end{bmatrix} \tag{8.19}$$

The matrix $\mathbf{X}$ is sometimes called *design matrix*, and equations that can be written in the form (8.18) (where the parameters enter the equation only linearly) are called *linear regressions*. In Python we can solve such problems in the same way as before:

```
X = np.column_stack( [np.ones_like(x), x] )
p_estimator = np.linalg.pinv(X) @ y
```

The most important feature of this type of model is that the parameters $\beta_i$ that we are looking for enter the model only linearly.

### 8.4.3  Line-Fit

For a line-fit, Eq. 8.19 is reduced to

$$
\begin{bmatrix} y_1 \\ y_2 \\ y_3 \\ \vdots \end{bmatrix} = \begin{bmatrix} x_1 & 1 \\ x_2 & 1 \\ x_3 & 1 \\ \vdots & \vdots \end{bmatrix} \cdot \begin{bmatrix} m \\ b \end{bmatrix}
$$

```python
# Generate a noisy line, with an slope of 0.5
# and a y-intercept of -30
x = np.arange(100)
y = -30 + 0.5*x + 2.5*np.random.randn(len(x))

# Plot the line
plt.plot(x, y, '.')
plt.axhline(0, LineStyle='--')

# Calculate the best-fit line
M = np.column_stack( [x, np.ones_like(x)] )
p_estimator = np.linalg.pinv(M) @ y

slope = p_estimator[0]          # "m"
y_intercept = p_estimator[1]    # "c"
```

Note that the sequence of columns in the design matrix determine the sequence of parameters in the parameter vector.

An alternative way to fit a line would be to see the line as a $1^{st}$ order polynomial, and use the function `np.polyfit` which can be used to fit polynomials of arbitrary order:

```python
p = np.polyfit(x,y,1)
y_fit = np.polyval(p, x)
```

If we are "only" interested in a line-fit, probably the most useful function is

```python
from scipy import stats
slope, intercept, r_value, p_value, std_err = stats.linregress(x, y)
```

because this provides additional useful information about the line-fit (for the correlation coefficient `r_value` see Sect. 8.1.1).

### 8.4.4  Polynomial-Fit

We can use the same approach to fit a polynomial curve to the data. For example, for a quadratic relationship between x and y is given by

$$
y = a * x^2 + b * x + c \tag{8.20}
$$

Written in matrix form this gives

$$
\begin{bmatrix} y_1 \\ y_2 \\ y_3 \\ \vdots \end{bmatrix} = \begin{bmatrix} x_1^2 & x_1 & 1 \\ x_2^2 & x_2 & 1 \\ x_3^2 & x_3 & 1 \\ \vdots & \vdots & \vdots \end{bmatrix} \cdot \begin{bmatrix} a \\ b \\ c \end{bmatrix} \tag{8.21}
$$

With this the problem has the form $\mathbf{y} \approx \mathbf{X} \cdot \boldsymbol{\beta}$, and can be solved in the same way as above. Or in other words:

> We can use a `linear` model to fit a quadratic (or higher order) polynomial, because the model parameters enter the equation only in a linear way.

### 8.4.5 Sine-Fit

For a sinusoidal oscillation where the frequency $\omega$ is known and where amplitude, phase delay, and offset are to be determined, the defining function is

$$x = offset \cdot 1 + amplitude \cdot \sin(\omega t + \delta) \tag{8.22}$$

Note that here *offset* and *amplitude* appear as linear parameters - but that the phase $\delta$ does not. This can be changed to a purely linear equation by expressing $sin(\omega * t + \delta)$ as the sum of a sine- and a cosine-wave:

$$x = offset \cdot 1 + a \cdot \sin(\omega t) + b \cdot \cos(\omega t) \tag{8.23}$$

From these parameters we can find amplitude and phase:

$$amplitude = \sqrt{a^2 + b^2} \tag{8.24}$$

$$\delta = \tan^{-1}\left(\frac{b}{a}\right) \tag{8.25}$$

In Python Eq. 8.25 should be implemented using `np.arctan2`, since `np.arctan2` always chooses the quadrant correctly (in contrast to `np.arctan`):

```
import numpy as np
np.rad2deg( np.arctan2(-0.1, -1) )
>>> -174.289

np.rad2deg( np.arctan(-0.1/-1) )
>>> 5.711
```

Now all the parameters that appear in the relationship Eq. 8.23 are linear, and the design matrix $\mathbf{X}$ can be written as (Fig. 8.7):

$$\mathbf{X} = \begin{bmatrix} 1 & \sin(\omega \cdot t_1) & \cos(\omega \cdot t_1) \\ 1 & \sin(\omega \cdot t_2) & \cos(\omega \cdot t_2) \\ 1 & \sin(\omega \cdot t_3) & \cos(\omega \cdot t_3) \\ \vdots & \vdots & \vdots \end{bmatrix} \tag{8.26}$$

The parameter vector is $\boldsymbol{\beta} = \begin{bmatrix} offset \\ a \\ b \end{bmatrix}$.

---

Listing 8.1: sine_fit.py

```
""" Demonstrate a fit to sinusoidal data """

# Import the standard packages
import numpy as np
import matplotlib.pyplot as plt

# Set the parameters for the sine-wave
freq = 0.5
offset = 3
```

```
delta = np.deg2rad(45)
amplitude = 2
rate = 10
duration = 3 * np.pi
omega = 2 * np.pi * freq

# Time
dt = 1/rate
t = np.arange(0,duration, dt)

# Simulate a noisy sine-wave
np.random.seed(123)                 # to make the noise reproducible
y = offset + amplitude * np.sin(omega*t + delta) + np.random.randn(len(t))

# Show the data
plt.plot(t, y, '--', label='noisy data')
plt.autoscale(axis='x', tight=True)

# Fit the data
M = np.column_stack( (np.ones(len(t)), np.sin(omega*t), np.cos(omega*t)) )
p = np.linalg.pinv(M)@y

# Extract the coefficients from the fit
found = {}
found['offset'] = p[0]
found['amp'] = np.sqrt(p[1]**2 + p[2]**2)
found['delta'] = np.rad2deg(np.arctan2(p[2], p[1]))
found['y'] = found['offset'] \
             + found['amp'] * np.sin(omega*t + np.deg2rad(found['delta']))

# Superpose the fit over the data
plt.plot(t, found['y'], label='fit', lw=3)

plt.legend(loc='lower right')
plt.axhline(ls='dotted')
plt.xlabel('Time [sec]')
plt.ylabel('Signal')

# Save and show the figure
out_file = 'sine_fit.jpg'
plt.savefig(out_file, dpi=200, quality=90)
print(f'Image saved to {out_file}')

plt.show()
```

### 8.4.6   Circle-Fit

The same concept can surprisingly be extended to find the best-fit circles to data. To do so the general equation for a circle has to be re-arranged to ensure we have a linear equation for the design matrix:

$$
\begin{aligned}
(x - x_c)^2 + (y - y_c)^2 &= r^2 \\
x^2 - 2xx_c + x_c^2 + y^2 - 2yy_c + y_c^2 &= r^2 \\
2x \cdot x_c + 2y \cdot y_c + 1 \cdot (r^2 - x_c^2 - y_c^2) &= x^2 + y^2
\end{aligned}
\tag{8.27}
$$

where $(x_c, y_c)$ indicate the circle center.
This gives $(x^2 + y^2) = \mathbf{X} \cdot \boldsymbol{\beta}$ , with

$$
\mathbf{X} = \begin{bmatrix} 1 & 2x_1 & 2y_1 \\ 1 & 2x_2 & 2y_2 \\ \vdots & \vdots & \vdots \end{bmatrix}
$$

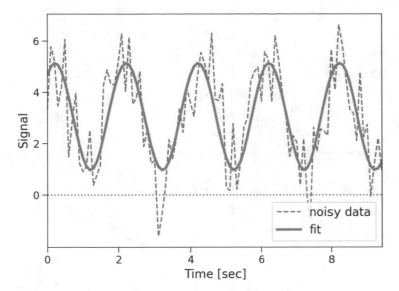

Figure 8.7: Sine-Fit (from Listing 8.1).

On the left hand side, $x^2 + y^2$ is known; and on the right hand side we again have a linear relationship! From this we get

$$x_c = \beta_1$$
$$y_c = \beta_2$$
$$r = \sqrt{\beta_0 + x_c^2 + y_c^2}$$

## 8.5 Confidence Intervals

As has been shown in Sect. 7.2.2, the *standard error of the mean* characterizes our uncertainty in the mean value, and the $\alpha\%$-confidence intervals (CIs) are the range that contains the true mean value with $\alpha\%$ probability. In this section we generalize this concept to the best-fit estimators for linear fits, and add some more information about the basic assumptions underlying the calculation of confidence intervals. In the previous section, we saw that it is possible to describe a linear trend via the least squares estimators. However, measured data always have variability and errors superimposed on the basic trend in which we are interested. These errors lead to uncertainty in the estimators.

### 8.5.1 Finding Confidence Intervals

To quantify the uncertainty in the best-fit estimators we have to make assumptions about the noise in the data. Some commonly made assumptions are that:

- the independent variables $x_1, ..., x_k$ are known exactly

- the residuals are roughly normally distributed

- the noise is only in the dependent variable y

- the residuals are independent of the values of $x_1, ..., x_k$

If any of these assumptions does not hold, then the fit does not contain all the information that the data can provide and may not be completely reliable.

Examples from cases where these assumptions do not hold is the *Anscombe's Quartet* (Fig. 8.8).

For the calculation of confidence intervals of the best-fit parameters it is best to resort to the package statsmodels, a Python package that provides classes and functions for the estimation

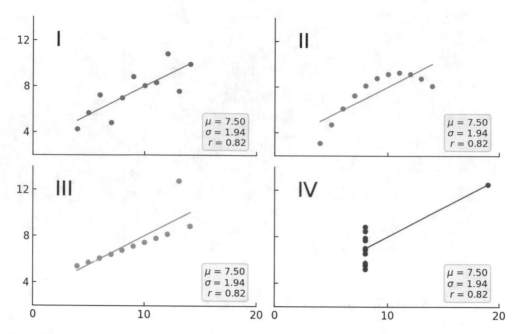

Figure 8.8: The sets in "Anscombes quartet" all have the same linear regression line but are themselves very different.

of many different statistical models and for statistical data exploration, and to make use of *pandas DataFrames*. `Statsmodels` uses the Python package `patsy` for describing statistical models, which allows to formulate "y is a function of x" simply as "$y \sim x$", where an offset is by default implicitly defined.

Listing 8.2: confidence_intervals.py

```
""" Calculation of confidence intervals for a line fit"""

# Import the required packages
import numpy as np
import pandas as pd
import statsmodels.formula.api as smf
from pprint import pprint

# Gnerate the data
np.random.seed(123)
x = np.arange(100)
y = 30 + 0.4*x + 5*np.random.randn(len(x))

# Use a "DataFrame" to contain the data, and a "formula" to define the function
df = pd.DataFrame({'x':x, 'y':y})
formula = 'y~x'
results = smf.ols(formula, data=df).fit()

# Show the results, with the default 95% confidence intervals
print(results.summary2())

# Address some of the fit-parameters
fit = {}
fit['parameters'] = results.params
fit['standard_error'] = results.bse
fit['ci_95'] = results.conf_int()                   # 95%   confidence intervals
fit['ci_999'] = results.conf_int(alpha=0.001)   # 99.9% confidence intervals

for key in [key for key in fit.keys() if key.startswith('ci')]:
    print(f'{key}: {fit[key]}')
```

Listing 8.2 uses the Python package *statsmodels* (https://www.statsmodels.org) to produce output that contains not only the model parameters of the line fit, ie. slope and intersection, and their corresponding probabilities and confidence intervals (middle section), but also information about the model itself (top section) and the normal distribution of the values (bottom section, down to "Condition No"). (For details on the individual parameters see Haslwanter, 2016).:

```
                    Results: Ordinary least squares
=================================================================
Model:                OLS           Adj. R-squared:       0.806
Dependent Variable:   y             AIC:                  633.8019
Date:                 2020-05-17 11:41 BIC:               639.0122
No. Observations:     100           Log-Likelihood:       -314.90
Df Model:             1             F-statistic:          413.5
Df Residuals:         98            Prob (F-statistic):   6.12e-37
R-squared:            0.808         Scale:                32.471
-----------------------------------------------------------------
                Coef.    Std.Err.     t     P>|t|    [0.025   0.975]
-----------------------------------------------------------------
Intercept       30.0654  1.1312   26.5790  0.0000  27.8206  32.3101
x                0.4014  0.0197   20.3347  0.0000   0.3622   0.4406
-----------------------------------------------------------------
Omnibus:                2.753         Durbin-Watson:        1.975
Prob(Omnibus):          0.252         Jarque-Bera (JB):     1.746
Skew:                   0.035         Prob(JB):             0.418
Kurtosis:               2.356         Condition No.:        114
=================================================================

ci_95:              0            1
Intercept   27.820605    32.310147
x            0.362243     0.440592
ci_999:             0            1
Intercept   26.227781    33.902972
x            0.334446     0.468389
```

For example, the code above produces a value of 0.4014 for the best-fit slope and a 95%-confidence interval of $[0.3622, 0.4406]$. This means that the slope of the best fit slope (i.e., $m$) is 0.40, and that the true slope lies with 95% probability between 0.3622 and 0.4406. (See also Fig. 8.9.)

The additional output from the program also indicates that the confidence interval widens as the required confidence level increases. For example, the 99.9% CIs ($[0.33, 0.47]$) are wider than the 95% CIs.

### 8.5.2   Confidence Intervals and Hypothesis Tests

A confidence interval for a parameter can be used directly to test the corresponding null hypothesis. For confidence intervals from parameters of regression models, the null hypothesis is:

*"The parameter is not significantly different from zero."*

If the confidence interval contains zero, then it is not possible to reject the null hypothesis at the corresponding confidence level. In other words, in that case there is no statistical evidence to suggest that the parameter is significantly different from zero.

### 8.5.3   Significance

If the upper *and* the lower limit of the 95%-confidence interval for the slope are both larger than 0, we speak of a *significant increase* in the data (e.g. Fig. 8.9). If in addition both limits of the 99.9% confidence intervals are above 0, we call this a *highly significant increase*. Similarly,

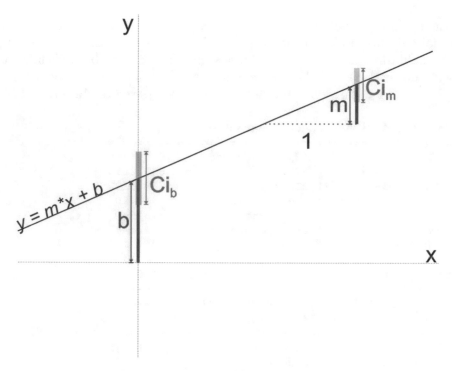

Figure 8.9: Confidence intervals (CIs) for offset b and for slope m.

if both limits of the confidence intervals are below zero, the data show a *significant* or a *highly significant decrease*. On the other hand, if the confidence intervals overlap zero we cannot claim that our data are in- or decreasing, even if the slope is positive or negative, respectively. The `pvalue` of the function `scipy.stats.linregress` provides the probability that the true slope is zero.

## 8.6    Fitting Nonlinear Functions

So far we only linear relationships between data and parameters have been considered. Note that even tasks such as fitting a sine-function with an offset and a phase can be expressed with linear relationships (see Sect. 8.4.5)! However, often the relationship will be visibly nonlinear and cannot be fit with a linear model. In that case it is necessary to select a *nonlinear model* for the data, and then attempt to fit the parameters of that model.

For example, for the data in Fig. 8.10 the falling part of the curve might reflect some physical process that is naturally modeled with an exponential decay. Since an exponential decay is fully defined by the starting-value and -time, the value of the asymptotic minimum, and the half-life, one way of fitting this curve would be to estimate these parameters independently. For example, the *offset* is approximately the last point in time, and the decay time is approximately given by the time it takes to decay from the $maxVal$ to $offset + (maxVal - offset) * e^{-1}$ .

Alternatively, one could use more sophisticated methods that attempt to estimate all of the model parameters exactly and simultaneously. For example `scipy.optimize.least_squares` provides powerful non-linear fits.

**Tips:**

- Nonlinear fits are more efficient and accurate when the user provides a decent estimate of the parameters as a starting point.

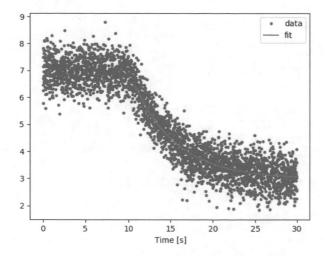

Figure 8.10: Plot of a variable that decays over time (from Listing 8.3).

- Choose a model/function that is likely to fit the data well, and—if you have a choice—reduce the number of parameters to be estimated.

## Example: Exponential Decay To a Constant Value

For fitting an exponential decay first the start of the exponential decay has to be determined. If the data are not too noisy, this can be automated by finding e.g. the first value *clearly* below the maximum.

For example, Listing 8.3 automatically fits an exponential decay to an offset (Fig. 8.10):

Listing 8.3: nonlinear_fit.py

```python
""" Demonstration of a non-linear fit """

# Import the standard packages
import numpy as np
import matplotlib.pyplot as plt
from scipy import optimize


def find_start(t, x):
    """ Find where the exponential decay approximately starts """
    max_val = np.max(x)
    level = 0.6      # 40% below maximum
    threshold = max_val * level
    first_below = np.min(np.where(x<threshold)[0])
    return t[first_below]


def model(x, t):
    """ Returns the residuals of the expected function """
    return x[0] + x[1] * np.exp(-t/x[2])


def err_fun(x, t, y):
    """ Error-function that is minimized by the fit """
    return model(x,t) - y
```

```
if __name__ == '__main__':

    # Define the parameters
    offset, amp, tau = 3, 4, 5
    t0, noise_amp = 10, 0.5

    # Generate the delayed noisy exponential decay
    time = np.arange(0, 30, 0.01)
    values = offset + amp*np.exp(-(time-t0)/tau)
    values[time<t0] = offset + amp
    np.random.seed(123)
    values += noise_amp * np.random.randn(len(time))

    # Fit the model
    t_start = find_start(time, values)
    decay = time > t_start
    x0 = [0, 1, 1]   # initial values for the fit
    par = optimize.least_squares(err_fun, x0, args = (time[decay],
            values[decay]))
    print(f'Fitted parameters: {par.x}')

    plt.plot(time, values, '.', label='data')
    plt.plot(time[decay], model(par.x, time[decay]), label='fit')
    plt.legend()
    plt.xlabel('Time [s]')
    plt.show()
```

## 8.7  Exercises

1. **Line Fits**

   - Take the data from `.\data\co2_mm_mlo.txt` and plot the `Monthly CO2-level` vs `Year`. (The data are from the *Earth System Research Laboratory* by NOAA, and were recorded on Mouna Loa.[1])

   - Fit a line to the CO2 levels vs year using `polyfit`.

   - Fit a quadratic curve to the same data using `polyfit`. What changes when you use `year-2000` instead of `year`? What causes this change in behavior?

   - Plot original data, line, and quadratic curve.

2. **Confidence Intervals** Use the same data as in Exercise 1, but now also determine the 95% confidence intervals. Answer the following 2 questions:

   - For the line-fit, is the slope of the line significantly rising?

   - For the quadratic curve-fit, is the quadratic contribution significant?

   - Does the addition of a cubic term improve the quality of the fit?

   **Notes:**

   - The order of the fitted polynomial should be as low as possible.

   - The order of the fit is too low if the residuals display a systematic pattern (see next exercise).

   - When the coefficient of the highest fitted power is not significant, the order is too high.

---

[1] https://www.esrl.noaa.gov/gmd/ccgg/trends

- When the highest order term is determined, then all lower order terms are also included. If for instance the *cubic* term is the highest significant term, then we would *retain all terms of smaller order than cubic.*

3. **Residuals**

   - Calculate the residuals, using the quadratic fit.
   - Plot the residuals as a function of time.
   - Select a recent time range where the residuals are approximately similar for each year.
   - Plot the residuals during that period.

4. **Sine Fit** Using the "similar residuals" from the previous exercise, and write a function that fits an offset sine-wave to those data.
   **Tip:** How do you create a sine-wave where one cycle is 12 points long?

# Chapter 9

# Spectral Signal Analysis

One of the most powerful tools for data analysis are integral transformations, such as the *Fourier transform* or the *Laplace transform*. Those transformations give us the ability to look at our data in a completely new way. At the same time, the varying formats in which they are presented, and the frequent use of complex exponentials, have needlessly daunted scores of aspiring students. The introductions to the Fourier transform in this chapter, and to the Laplace transform in the next one, are intended to provide an intuitive understanding of the principles behind those transformations and demonstrations on how to use them with Python. They cover by no means all aspects of those transformations. An excellent introduction to the mathematical background and to applications of the Fourier transform are the books by Julius O. Smith III (e.g. Smith (2007)). And Python implementations of many aspects of the Fourier transform are well demonstrated in Unpingco (2014).

## 9.1 Transforming Data

The Fourier transform can transform signals from the time-domain to the frequency-domain. Two sample signals will be used to demonstrate the Fourier Transforms: a simulated signal, with three well defined frequency contributions; and a real sound signal.

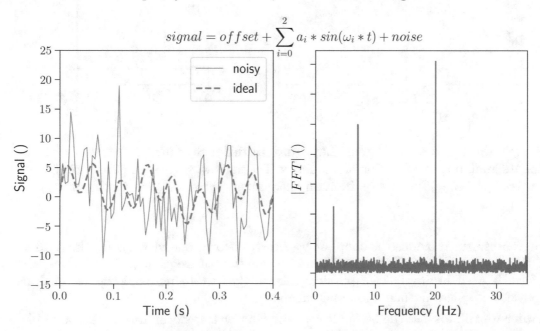

Figure 9.1: Sample data 1: a simulated signal, with a well defined frequency content. Time-view (left) and frequency-view (right). Arbitrary units are indicated in axis labels with "()".

T. Haslwanter, *An Introduction to Hands-on Signal Analysis with Python*, https://doi.org/10.1007/978-3-030-57903-6_9

The first, simulated signal is shown in Fig. 9.1: viewed as a function of time (left panel) the signal looks fairly random. But looking at the frequency content (right panel) the picture is pretty clear: only very few frequency components contribute to the signal.

The second data sample that will be used in this chapter is a real sound signal: the left panel in Fig. 9.2 shows the air pressure of a sound wave as a function of time. Zooming in (middle) we can see some repetitive pattern, but it is still not clear what is happening. However, looking at the same data as a function of frequency (Fig. 9.2, right) gives a much simpler picture: the simple, regular structure of the signal as a function of frequency indicates that this signal might be a sound from a string instrument, the frequency content of which typically consists of multiples of a base frequency. In this example I have hit a key on a piano, producing a tone with 440 Hz. (The relative intensity of multiples of this frequency, the so-called "higher harmonics", depend on the type of instrument.)

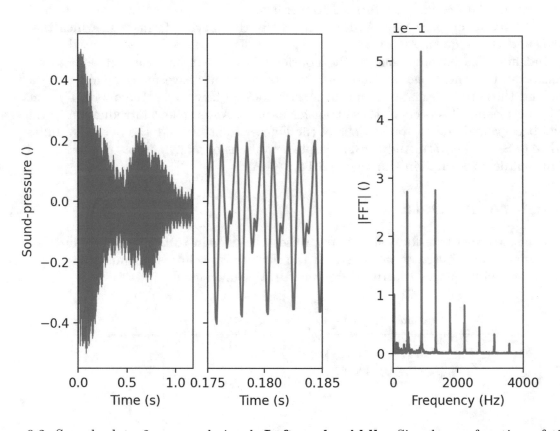

Figure 9.2: Sample data 2: a sound signal. **Left and middle:** Signal as a function of time. **Right:** The same signal, but as a function of frequency. The narrow, regularly spaced amplitudes indicate that this is probably a pure tone from an instrument. (From CQ `F9_fft_sound.py`).

The Fourier transform provides the simplest way to transform one view (here, the data as a function of time) into another (here, the data as a function of frequency). In practice, the Fourier transform is used in any field of physical science that uses sinusoidal signals, such as engineering, physics, applied mathematics, and chemistry.

In this chapter we will discuss the basic ideas of the Fourier transform, how it can be implemented in Python, and how the results can be interpreted. A very good extensive introduction can be found for example in the book by Smith (2007). An excellent shorter overview of Fourier analysis is the article Harris (1998).

The basic idea of the Fourier transform is presented in Fig. 9.3: every digitally recorded signal can be represented as a weighted sum of pure oscillations, with amplitude and phase

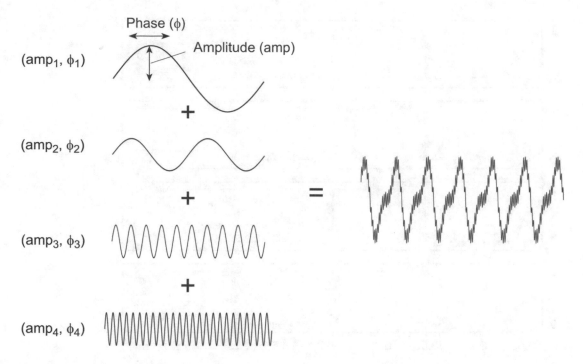

Figure 9.3: Construction of a function as a sum of oscillations, with the amplitude and phase of each oscillation appropriately adjusted. For a given signal, the *Fourier transform* provides the (amplitude/phase)-information for each oscillation.

of each oscillation properly adjusted[1]. When the correct combination of sine waves are all added together, the end result is identical to the function of interest. The required frequency, amplitude and phase of each oscillation can be found with the *Fourier transform*. In practise, a small number of frequencies is often sufficient to characterize the main features of the original signal (see Fig. 9.4).

## 9.2 Fourier Integral

### 9.2.1 Definition and Interpretation

To transform a continuous function from the time domain to the frequency domain, one uses the *Fourier Integral*. A common convention is to indicate functions of time with lowercase variables, and functions of frequency with uppercase:

$$X(f) = \int_{-\infty}^{\infty} x(t)e^{-j2\pi ft}dt \qquad (9.1)$$

where $f$ represents the frequency. The inverse transform is given by

$$x(t) = \int_{-\infty}^{\infty} X(k)e^{j2\pi ft}df \qquad (9.2)$$

The Fourier transform $X(f)$ is a complex number and has a straightforward, intuitive interpretation: its magnitude is the amplitude of the corresponding frequency $f$, and its phase the phase-shift of the corresponding frequency (see Fig. 5.12, and compare it to Fig. 9.3, top).

**Note:** Some authors choose to define the Fourier transform (and its inverse) in terms of the angular frequency $\omega = 2\pi f$. In this case a scaling factor is required that is in total $\frac{1}{2\pi}$. The $\frac{1}{2\pi}$

---

[1]Here many details, which arise when investigating signals that are not continuous and/or infinite, are glossed over. But since those details do not affect the finite discrete signals that are used for digital signal analysis, all those details are skipped here.

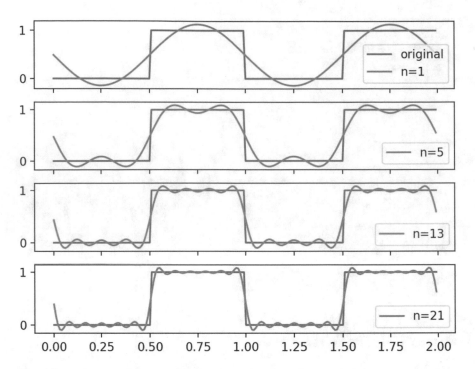

Figure 9.4: One important property of Fourier Transform: in practice, a fairly small number of dominant frequencies is often sufficient to characterize the main features of the underlying signal.

may scale the forward transform or the inverse transform, or $\frac{1}{\sqrt{2\pi}}$ may scale both (*Symmetrical Fourier Transform*).

### 9.2.2    Complex Exponential Notation

Oscillations can be formulated in three different but equivalent ways:

1. $osc(t) = (r * e^{j\phi}) * e^{j2\pi ft}$

2. $osc(t) = r * \sin(2\pi ft + \phi)$

3. $osc(t) = a * \cos(2\pi ft) + b * \sin(2\pi ft)$

The connection between the three comes from Euler's formula (Eq. 1.5), which states that

$$e^{j2\pi ft} = \cos(2\pi ft) + j\sin(2\pi ft)$$

From this it follows that sine and cosine waves can be expressed in terms of $e^{j2\pi ft}$ and $e^{-j2\pi ft}$ (see also Fig. 9.5):

$$\cos(2\pi ft) = \frac{1}{2}\left(e^{j2\pi ft} + e^{-j2\pi ft}\right) \qquad (9.3)$$

$$\sin(2\pi ft) = \frac{1}{2j}\left(e^{j2\pi ft} - e^{-j2\pi ft}\right)$$

Since

$$(r * e^{j\phi}) * e^{j2\pi ft} = r * e^{j(2\pi ft + \phi)} \qquad (9.4)$$

a multiplication of $e^{j2\pi ft}$ with a complex number corresponds to scaling and phase-shifting an oscillation.

One final mathematical property of trigonometric functions that is required here is:

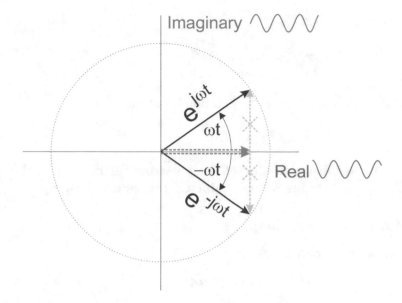

Figure 9.5: To cancel the imaginary components of an exponential oscillation $e^{j\omega t}$, the values can be summed with the corresponding values for negative frequencies $e^{-j\omega t}$.

- Any combination of a sine and a cosine wave with frequency k is again a sinusoid with this frequency. The amplitude and phase of the oscillation depend on the relative components of the sine and the cosine contribution.

Specifically,

$$a * \cos(2\pi ft) + b * \sin(2\pi ft) = c * \sin(2\pi ft + \phi) \tag{9.5}$$

where $c = \sqrt{a^2 + b^2}$ , and $\phi = \tan^{-1}(b/a)$.

As a consequence of these relationships, scaled and phase-shifted sine-waves can therefore be written either as

1. or an exponential oscillation function multiplied with a complex number,

2. a scaled, phase-shifted sine-signal,

3. or the sum of a sine- and a cosine signal,

corresponding to the equations at the beginning of this section. These three representations are completely equivalent. Since the most compact form is the exponential representation, we will use it in the following.

### 9.2.3   Examples

To demonstrate two applications of the Fourier transform, we show which frequency contributions a constant offset contains, and which frequencies are needed to generate a pure sine or cosine wave.

**Fourier Transform of a Constant**

Let us start with the constant function

$$x(t) = 1 . \tag{9.6}$$

The Fourier transform of Eq. 9.6 is

$$X(f) = \int_{-\infty}^{\infty} e^{-j2\pi ft} dt$$

This integral turns out to be the *Dirac delta function*[2] centered at 0:

$$X(f) = \delta(0)$$

If you feel uncomfortable working with an "infinitely narrow" function, just think of it as an impulse at t=0. That captures the important characteristics here sufficiently.

**Fourier Transform of a Pure Oscillation**

Next we take a simple cosine wave with frequency $\nu$:

$$x(t) = A * \cos(2\pi\nu t)$$

The Fourier transform is

$$X(f) = \int_{-\infty}^{\infty} A * \cos(2\pi\nu t)\, e^{-j2\pi ft} dt$$

Using the complex exponential representation, this becomes

$$X(f) = \int_{-\infty}^{\infty} A * \left( \frac{e^{j2\pi\nu t} + e^{-j2\pi\nu t}}{2} \right) e^{-j2\pi ft} dt$$

$$= \frac{A}{2} * \int_{-\infty}^{\infty} \left( e^{-j2\pi(f-\nu)t} + e^{-j2\pi(f+\nu)t} \right) dt$$

Finally, we can write this in terms of Dirac delta functions:

$$X_{cosine}(f) = \frac{A}{2} \left[ \delta(f - \nu) + \delta(f + \nu) \right] \tag{9.7}$$

The result is illustrated in Fig. 9.5: the imaginary components of an exponential oscillation can be eliminated by adding the corresponding values from an oscillation with the negative frequency.

The same procedure for a sine-signal leads to

$$X_{sine}(f) = \frac{A}{2j} \left[ \delta(f - \nu) - \delta(f + \nu) \right] \tag{9.8}$$

## 9.3   Fourier Series

### 9.3.1   Definition

Real measurement signals are never infinite. They always have a beginning and an end. Using a little trick such signals can be turned into "periodic" infinite signals, simply by repeating the signal again and again (see Fig. 9.6).

It can be shown that every periodic function can be decomposed into a sum of sinusoids with frequencies that are multiples of a fundamental frequency, the so-called *Fourier Series*:

$$x(t) = a_0 + \sum_{n=1}^{\infty} \left[ a_n * \cos(2\pi n f_p t) + b_n * \sin(2\pi n f_p t) \right] \tag{9.9}$$

---

[2]A *Dirac delta function* is an infinitely narrow single peaked function, with a height such that the integral over the peak gives exactly one, and zero everywhere else.

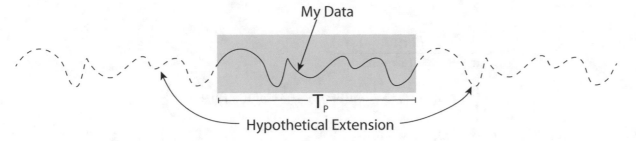

Figure 9.6: Artificially extended signal.

The fundamental frequency $f_p$ is determined by:

$$f_p = \frac{1}{T_P}$$

where $T_P$ is the duration of the data set, i.e., the duration of one "period". Again, the Fourier series can be written more succinctly in complex notation

$$x(t) = \sum_{n=-\infty}^{\infty} X_n * e^{j2\pi n f_p t} = \sum_{n=-\infty}^{\infty} X_n * e^{j2\pi n \frac{t}{T_P}} \tag{9.10}$$

where the values of $X_n$ are given by

$$X_n = \frac{1}{T_P} \int_{\tau}^{\tau + T_P} x(t) * e^{-j2\pi n \frac{t}{T_P}} dt \tag{9.11}$$

### 9.3.2 Applications

This periodic repetition of time-limited signals has important consequences, especially for short-duration signals. Take for examples two short signals, both containing a jump from 0 to 1. But while the first signal jumps in the middle of the signal (Fig. 9.7A, top), the second jumps just before the end of the signal (Fig. 9.7A, bottom). Repetition of the signals leads in the first case to a step-signal ((Fig. 9.7B, top), and in the second case to a sequence of impulses (Fig. 9.7B, bottom). (A very good introduction on how to deal with these effects has been given by Harris (1998).)

This property can also be formulated differently: edge effects are especially important for short duration signals.

**Square Wave:** Using Fourier expansion with cycle frequency $f$ over time $t$, we can represent for example an ideal square wave with a peak to peak amplitude of 2 as an infinite series of the

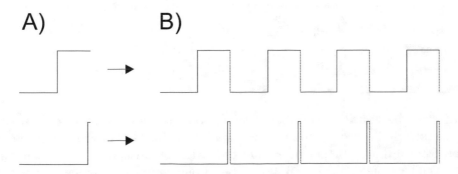

Figure 9.7: Effects of signal repetition for short duration signals.

form

$$x_{\text{square}}(t) = \frac{4}{\pi} \sum_{k=1}^{\infty} \frac{\sin(2\pi(2k-1)ft)}{(2k-1)}$$

$$= \frac{4}{\pi} \left( \sin(2\pi ft) + \frac{1}{3}\sin(6\pi ft) + \frac{1}{5}\sin(10\pi ft) + \cdots \right) \tag{9.12}$$

This corresponds to the example shown in Fig. 9.4.

## 9.4  Discrete/Fast Fourier Transform

### Discrete Fourier Transform (DFT)

In practice, when working with digital representations of data sampled from a continuously varying signal, the data sets are not only time-limited but also discrete. For such data sets only a finite number of sinusoids are required to represent the signal. Specifically, for $N$ discrete data points only $N$ sinusoids are required to make up the signal (Fig. 9.8).

If $N$ data points were sampled with a constant sampling frequency, then the Fourier coefficients $F_n$ can be obtained with

$$X_n = \sum_{\tau=0}^{N-1} x_\tau * e^{-j2\pi \frac{n \cdot \tau}{N}} \quad \text{with } n = 0, \cdots, N-1. \tag{9.13}$$

The inverse Fourier transform is given by

$$x_\tau = \frac{1}{N} \sum_{n=0}^{N-1} X_n * e^{j2\pi \frac{n \cdot \tau}{N}} \quad \text{with } \tau = 0, \cdots, N-1. \tag{9.14}$$

Since these equations contain a limited number of discrete data points (Eq. 9.13) and a limited number of discrete waves (Eq. 9.14), this transform is referred to as *Discrete Fourier Transform (DFT)*.

**Conventions**   There are many ways to define the DFT, varying in the sign of the exponent, normalization, etc. The definition of DFT used here is the same as the one used by Smith (2007), cited in Wikipedia, and implemented in *numpy*. However, using this definition the Fourier

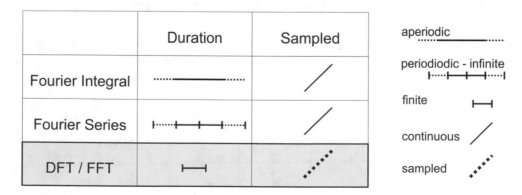

Figure 9.8: The choice of the Fourier Transform depends on the characteristics of the signal. Real measurement signals (indicated by the shaded area) are always finite and discrete, and can be reconstructed from a finite number of oscillations with the *Discrete Fourier Transform (DFT)*. If the signal length equals $2^n$, the DFT can be implemented very efficiently with the algorithms of the *Fast Fourier Transform (FFT)*.

coefficients depend on the sample frequency. For example, using this definition the largest Fourier coefficient for one cycle of a pure sine wave corresponds to half the number of sample points

```
import numpy as np
num_pts = 100
t = np.arange(0, 1, 1/num_pts)
x = np.sin(2*np.pi*t)
fft_coeffs = np.fft.fft(x)
print(np.max(np.abs(fft_coeffs)))
>>> 50
```

For this reason, some authors prefer the convention that the "1/N" appears in Eq. 9.13 instead of Eq. 9.14.

## Fast Fourier Transform (FFT)

The DFT has become a mainstay of numerical computing in part because of a very fast algorithm for computing it, called the "Fast Fourier Transform (FFT)". That algorithm was already known to Gauss (1805) (Heideman 1984), and was brought to light in its current form by Cooley and Tukey (1965). If the number of data points is exactly $N = 2^n$, the number of multiplications required can be reduced by many orders of magnitude, especially for long signals. Press et al. (2007) provide an accessible introduction to the FFT and its applications.

Because the discrete Fourier transform separates its input into components that contribute at discrete frequencies, it has a great number of applications in digital signal processing e.g. for filtering. In this context the discrete input to the transform is customarily referred to as a *signal*, which exists in the time domain. The output is called a *spectrum* or *transform* and exists in the frequency domain.

To make use of the speed benefits of the FFT, signals that contain less than $2^n$ data points are often *zero-padded*, i.e. extended with 0's until their length matches the next power of 2. Zero padding in the time domain is also used extensively in practice to compute heavily interpolated spectra by taking the DFT of the zero-padded signal.

## Real-Valued Signals

Measurement signals always consist of real-valued data and therefore constitute the most common input for Fourier transforms. For such signals the amplitudes of the spectrum show a surprising symmetry (see Fig. 9.9): the last Fourier coefficient is the complex conjugate (see Eq. 1.4) of the second one, the last-but-one the complex conjugate of the third one etc. (See the example on p. 168, bottom)

The reason behind this is that for real-valued signals the following equation must hold

$$X(f) = X(-f)^*$$  (9.15)

where "*" indicates the complex conjugate. In words, for real-valued signals the Fourier transform needs to include positive and negative frequency components of the same magnitude, so that the imaginary contributions of the two can cancel out. One example is given by the Fourier transform of a pure cosine oscillation in Eq. 9.7 (one a $\delta$-function at $k = \nu$, and a corresponding one at $k = -\nu$; see also Fig. 9.5). Another is the frequency spectrum in Fig. 9.9, which is also symmetrical about the center.

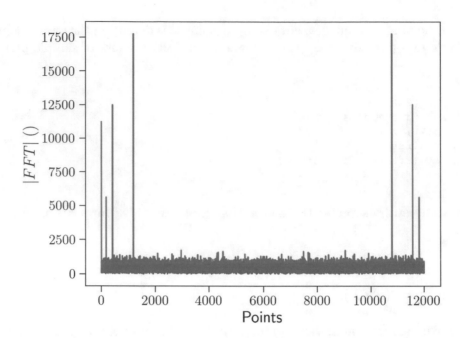

Figure 9.9: For real inputs, the magnitude of the Fourier spectrum is symmetric about the Nyquist frequency.

## Python Implementation: Noisy Sine-Waves

To demonstrate a practical application we generate a simulated data set, consisting a noisy superposition of three sine waves with an offset (see Fig. 9.1):

```python
import numpy as np

# First set the parameters
rate = 100        # [Hz]
duration = 60     # [sec]
freqs = [3, 7, 20]
amps = [1, 2, 3]
offset = 1
np.random.seed(12345)

# Then calculate the data
t = np.arange(0, duration, 1/rate)

sig = np.zeros_like(t)
for (amp, freq) in zip(amps, freqs):
    omega = 2 * np.pi * freq
    sig += amp * np.sin(omega*t)

# Add some noise, and an offset
sig += np.random.randn(len(sig))*5 + offset
```

In Python, the DFT of data set and the *power spectrum* (defined as the square of the amplitudes, see below) can be found via the numpy commands

```python
from numpy import fft

fourier = fft.fft(sig)
Pxx = np.abs(fourier)**2    # calculate power spectrum by hand
```

and the inverse transform with

```
sig_reconstructed = fft.ifft(fourier);
```

**Note:** The minute imaginary components that are generated by the inverse Fourier transform with `fft.ifft` are numerical artifacts.

## Frequencies

A detail frequently missed by novices is the question: "Which frequency corresponds to $F_n$?" If there are $N$ data points and the sampling period is $T_s$ , the $n^{th}$ frequency (in [Hz]) is given by

$$f_n = \frac{n}{N \cdot T_s}, \ 0 \leq n \leq N - 1 \tag{9.16}$$

Or in words:

- The first Fourier component corresponds to $e^{j0} = 1$, and is therefore proportional to the offset of the signal.

- The lowest oscillating frequency is $\frac{1}{N \cdot T_s}$ [Hz] , i.e. $\frac{1}{signal duration}$.

- For real valued data, the highest independent frequency is half the sampling frequency, $\frac{1}{2T_s}$ (Nyquist-Shannon theorem).

- For real valued data, the upper half of the Fourier coefficients have to be the complex conjugate of the lower half.

The simplest way to calculate the corresponding frequencies is the function

```
freqs = fft.fftfreq(len(sig), 1/rate)
```

This function also uses a mathematical "trick": since the exponent in the DFT (Eq. 9.14) is periodic in the frequency

$$e^{2\pi j \frac{-n \cdot \tau}{N}} = e^{2\pi j \frac{-n \cdot \tau}{N}} * 1 = e^{2\pi j \frac{-n \cdot \tau}{N}} * e^{2\pi j \frac{N \cdot \tau}{N}} = e^{2\pi j \frac{(-n+N) \cdot \tau}{N}} \tag{9.17}$$

frequencies above the Nyquist frequency can be interpreted as negative frequencies (Fig. 9.10).

## Single-Sided Spectrum

Since for real-values signals the negative frequency components are the complex conjugate of the corresponding positive frequencies (Eq. 9.15), those components can be left away as they contain no additional information. This is called *single-sided spectrum* (e.g. Fig. 9.2).

In Python computation of the one-dimensional DFT for real input can be obtained with the command `np.fft.rfft`, and the inverse transform with `np.fft.irfft`.

## Interpretation of Fourier Coefficients

The right panel of Fig. 9.1 shows the spectrum (i.e. the magnitude of the Fourier coefficients) of the simulated noisy sine-wave from above (duration=1 min, sample_rate=100 Hz). The first three Fourier components are:

```
fourier[:3]
>>>    [6027.0+0.j    -245.5-139.6j    34.4+56.3j]
```

What do we know about the data?

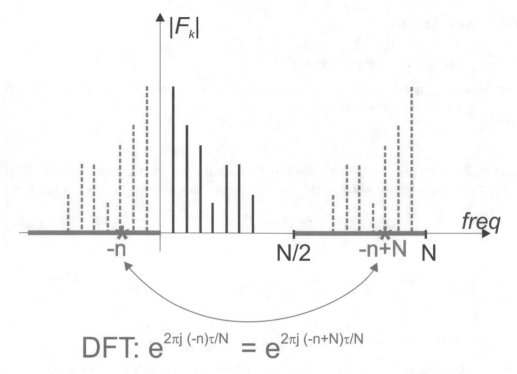

Figure 9.10: Due to the periodicity of $e^{2\pi jft}$ in $f$, frequencies above the Nyquist frequency ($N/2$) can be equivalently indicated as negative frequencies.

- Since `sig` contains *60sec * 100Hz = 6000* data points, `fourier` contains 6000 complex numbers.

- Since $e^0 = 1$, the first value is proportional to the offset, which is $\frac{6027}{6000} \approx 1$.

- The amplitude of the first frequency is $\sqrt{245.5^2 + 139.6^2} = 282.4$.

- The phase-shift of the first frequency is `theta = np.arctan2(-139.6, -245,5)` $\rightarrow -2.62$ rad . (Note that `np.arctan2` gives the correct value, even for angles $> 90$ deg. This is *not* the case with `np.arctan`!)

- The first frequency is determined by the length of the recording, and is here $f_1 = 1/60$ Hz.

- Analogous, (amplitude, phase-shift, frequency) of the second component are (*66.0, 1.02 rad*, 2/60 Hz respectively).

- Since the input values are real the highest frequency component which contains new information is the *Nyquist frequency*, which is there $100/2 = 50$ Hz. (Note that this no longer holds if the inputs values are complex!)

- Since the input values are real the values of the top two Fourier coefficients are also known: they have to be the complex conjugate of the lowest two oscillating contributions:

```
fourier[:3]
>>>    [6027.0+0.j     -245.5-139.6j     34.4+56.3j]
fourier[-2:]
>>>    [34.4-56.3j     -245.5+139.6j]
```

## 9.5 Spectral Density Estimation

For many applications the phase of the Fourier coefficients is not very important, and the main goal is to obtain an estimate for the spectral density of the signal, also know as the *power spectrum*. This spectral density characterizes the frequency content of the signal.

Depending on the application different methods can be used to find the power spectrum. They can be grouped into *non-parametric* and *parametric* spectral density estimation techniques. The most common non-parametric techniques, which we will discuss below, are the

**Periodogram** the modulus-squared of the discrete Fourier transform, and

**Welch's method** a windowed version of the periodogram that uses time-averaging.

Parametric techniques are the *autoregressive model (AR)*, the *moving-average model (MA)*, and the *autoregressive moving average (ARMA)*-model. For a more detailed description of those, see for example Fan and Yao (2003).

### 9.5.1 Periodogram

Application of the FFT to a time-dependent signal returns the complex Fourier coefficients $X_n$. The signal "power" contributed by each oscillation is proportional to the square of the amplitude of the Fourier coefficients:

$$P_n = F_n \cdot F_n{}^* = |F_n|^2 \tag{9.18}$$

This is called the *periodogram* or *power spectrum* of the signal, or since it describes the distribution of power – which is proportional to the square of the amplitude – over the individual frequency components composing that signal.

A convenient way to calculate the periodogram is the command

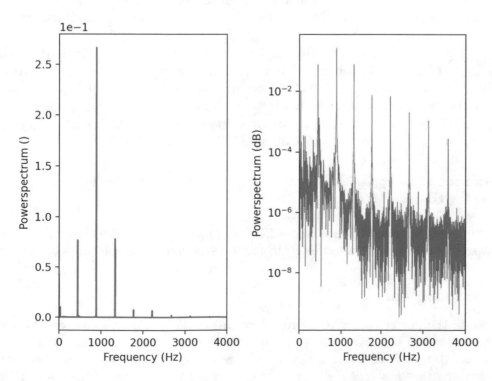

Figure 9.11: **Left:** The power spectrum of the signal in Fig. 9.2, plotted on a linear scale. **Right:** the same power spectrum, plotted with the ordinate on a logarithmic scale. Note that the higher harmonics are much better visible on the logarithmic scale. (From `F9_fft_sound.py`).

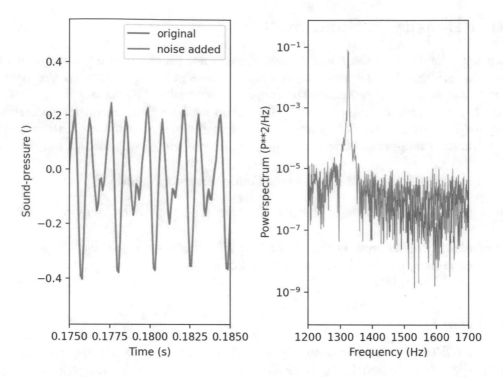

Figure 9.12: **Left:** Adding a tiny amount of noise does not change the signal noticeably. **Right:** In the corresponding power spectra the peaks are also only marginally affected. However, the noisy parts of the signal change drastically. This indicates that the relevant information is contained in the peaks of the power spectrum. (From `F9_fft_sound.py`).

```
f, Pxx = scipy.signal.periodogram(data, fs)
```

The spectral density in the periodogram is commonly displayed on a logarithmic scale, since this typically provides a better view of smaller but often important frequency contributions (see Fig. 9.11).

The peaks of the power spectrum contain the important frequency contributions to the signal. The noisy parts of the signal depend on numerical differences and on artifacts arising from the "windowing" used in the calculation of the power spectrum (see Sect. 9.7.1) and don't contribute significantly to the shape of the signal (see Fig. 9.12).

**Note:** The spectral density calculated with the command `scipy.signal.periodogram` contains two differences compared to the values obtained directly with Eq. 9.18:

- `signal.periodogram` applies a "window" win (see Fig. 9.15) to the whole signal, thereby removing the data offset.

- By default the result is scaled by `scale = 1.0 / (fs * (win*win).sum())`, producing the spectral density expressed as RMS (root-mean-square) amplitude per Hz.

### 9.5.2   Welch Periodogram

To reduce the noise in the power spectrum a procedure called *Welsh-Periodogram* can be employed. Thereby the data set is broken down into a number of separate segments, the power spectrum is calculated for each of them, and then the resulting power spectra are averaged (Fig. 9.13). The resulting power signal is returned (*Power Spectral Density - PSD*). Assuming that the frequency distribution stays approximately constant, this enhances the dominant components and reduces random noise. The cost is a reduction of the frequency resolution (see Fig. 9.14).

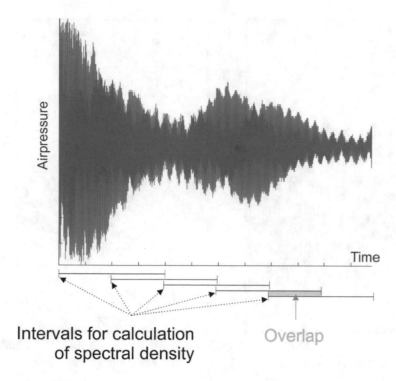

Figure 9.13: Principle of Welch's method for calculating the spectral power density: the PSD (*power spectral density*) is calculated for each interval indicated, and then all PSDs are averaged. If the segments are non-overlapping, the method is sometimes called *Bartlett's method.*

```
f_welch, P_welch = signal.welch(sig, fs)
```

🐍 python **Code:** F9_1_FFT_sines.py shows the calculation of the spectral density for the simulated signal in Fig. 9.1 with three different methods.

## 9.6   Fourier Transformation, Convolution, and Cross-Correlation

### 9.6.1   Convolution

For short signals convolutions can be calculated efficiently directly using Eq. 5.6. But for longer signals it becomes much more efficient to calculate it through the Fourier transform. The basis for this is given by the *convolution theorem*:

$$F\{f * g\} = \mathcal{F}\{f\} \cdot \mathcal{F}\{g\} \tag{9.19}$$

where $\mathcal{F}$ denotes the Fourier transform, $*$ a convolution, and $\cdot$ denotes point-wise multiplication.

By applying the inverse Fourier transform $\mathcal{F}^{-1}$ we can write:

$$f * g = \mathcal{F}^{-1}\big\{\mathcal{F}\{f\} \cdot \mathcal{F}\{g\}\big\} \tag{9.20}$$

This is a very efficient way to calculate a convolution. For example, for 2D images an FFT-based implementation of the convolution is the most efficient one for kernel sizes larger than $8 \times 8$ - $12 \times 12$, depending on the type of the implementation (linear vs. circular convolution).

Also remember that an application of an FIR-filter with filter coefficients **b** is equivalent to a convolution with a signal **b** (see Sect. 5.2.2).

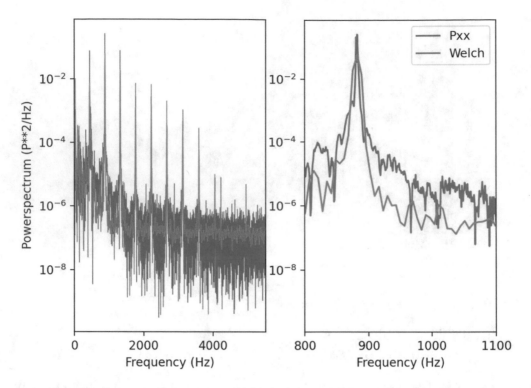

Figure 9.14: Different methods exist to reduce the noise in the power spectrum. (From `F9_fft_sound.py`).

### 9.6.2  Cross-Correlation

Since convolution and cross correlation are closely related, there is a corresponding relationship to Eq. 9.19, the *correlation theorem*:

$$Corr(g, h) \leftrightarrow G \cdot H^* \qquad (9.21)$$

where $g, h$ are functions of time, $G, H$ the corresponding Fourier transforms, and $H^*$ is the complex conjugate of $H$. If $h(t)$ is real, then $H(-f) = H^*(f)$.

**Note:** Which of the signals gets conjugated in Eq. 9.21 depends on the definition of cross correlation - and different definitions are in use!

🐍 python **Code:** `F9_FFT_sound.py` generates most of the figures in this Chapter.

## 9.7  Time Dependent Fourier Transform

### 9.7.1  Windowing

The Fourier transform calculates the frequency content of the whole signal. But often signals are changing their characteristics over time. In that case one also wants to know how the power spectrum changes with time. The simplest way is to take only a short segment of data at a time and calculate the corresponding power spectrum. This approach is called *Short Time Fourier Transform (STFT)*.

To obtain time-selective information from the Fourier transform, "windowing" can be applied to a signal. Figure 9.15 illustrates this principle. The top left panel shows the sound signal that we have been using so far. The bottom left shows a window that selects a short segment of that sound signal. The right panels show how the windowing is executed: the underlying signal is multiplied with the window, to obtain the signal during a short time period around the center

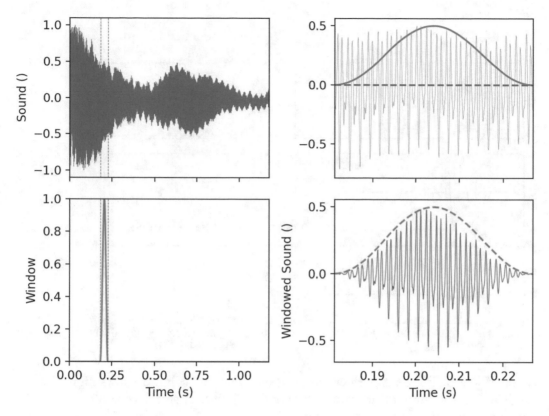

Figure 9.15: "Windowing" of a signal, to obtain time-dependent signal characteristics.

of that window. Calculating the spectral density during that time period provides the power spectrum during the time of the window.

For short signals edge effects can significantly distort the power spectrum of the signals, since we are assuming that our signal is periodic. Using windows that are tapered at the edges can eliminate such edge-artifacts, and can also be used to eliminate the offset in a signal. Figure 9.16 illustrates how clipping a purely periodic signal can introduce artifacts in the power spectrum (Fig. 9.16, top and middle). Tapered edges minimize the high-frequency ringing artifacts associated with hard edges, which are also called "spectral leakage". A *Hanning window* was used for Fig. 9.16, bottom. While a rectangular window provides better frequency resolution (Fig. 9.16, middle), the Hanning window exhibit fewer artifacts.

**Note:** Windowing and spectral leakage is a very important topic for engineers, and windowing of digitized signals can lead to surprising effects. It is dealt with in much more detail in specialized books, such as Smith (2007) or Unpingco (2014).

### 9.7.2 Example: Human Vowels

A good example for the application of STFT is the analysis of auditory signals. Figure 9.17. shows the STFT of a person speaking the vowels "a - e - i - o - u" (it is by a German speaker, so the vowels sound like in "Charlie - Echo - India - Oscar - Zulu", or "hut - hat - hit - hot - put"). At approximately 3, 6, 9, 12, and 16 sec, one can nicely see the harmonics of the regular vibrations of the vocal cords, which are the characteristic feature of vowels.

More information about sound processing is presented https://github.com/thomas-haslwanter/sapy.

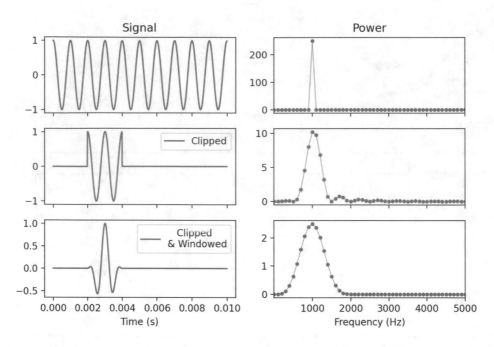

Figure 9.16: Effect of the STFT on a perfect cosine wave. The left column shows signals as a function of time. The right column shows the corresponding power spectra. In the bottom row the signal has been multiplied with a Hanning window.

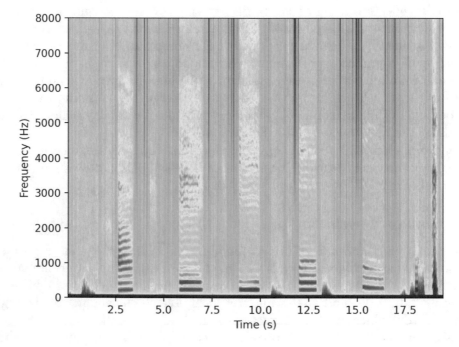

Figure 9.17: Spectrogram of the German vowels "a,e,i,o,u" from the sound file .\data\vowels.wav. These correspond approximately to the vowels in the English words "hut - hat - hit - hot - put". Calculated using the command plt.specgram(data, NFFT=NFFT, Fs=fs, noverlap=256, cmap=cm.jet).

## 9.8 Exercises

### 1. Power spectrum

The commands

```
# Get the required packages
import numpy as np
import matplotlib.pyplot as plt

# Generate the data
t = np.arange(0, 10, 0.1)
x = np.sin(t)
fft_data = np.fft.fft(x)
Pxx = np.abs(fft_data)**2

# Show a few values
print(fft_data[:3])

# Plot the signal
fig, axs = plt.subplots(1,2)
axs[0].plot(t,x)
axs[0].set_xlabel('Time (s)')
axs[0].set_ylabel('sin(t)')

axs[1].plot(Pxx, '-*')
axs[1].set_xlim([-0.2, 10.2])
axs[1].set_xlabel('Points')
axs[1].set_ylabel('Power()')

plt.tight_layout()
plt.show()
```

produce the following plot and output:

```
>>>  array([ 18.65 +0.j  ,  30.64 +5.65j, -31.5 -11.8j ])
```

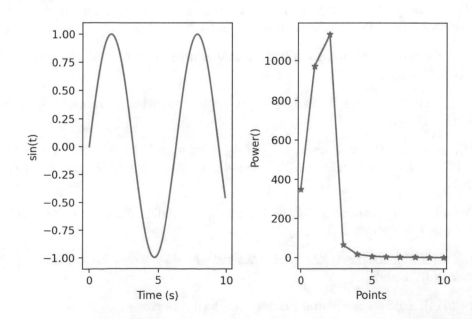

Answer the following questions:

- Why does the power-spectrum have more than one frequency component, although the data are from a perfect sine-wave?

- What would would you need to do to get a power-spectrum with only one frequency component?

- Why is the first value of the output "real" (18.65)

- How can we interpret the second and third Fourier coefficients (30.64 +5.65j, -31.5 -11.8j)?

- What are the frequencies corresponding to the individual "Points" in the right panel? Give the explicit values for the second and third Fourier coefficients.

**Hint:** See the discussion of time-limited signals in Sect. 9.3, and try to think about the consequences for the signal in this exercise.

2. **Hand-coded Fourier Transform**
   For

```
t = np.arange(0, 10, 0.1)
x = np.sin(t) + 3*np.cos(3*t)
```

calculate the DFT by hand from Eq. 9.14, and compare it to `np.fft.fft(x)`.

3. **Your Voice**

Install the open-source program *audacity* (https://www.audacityteam.org) on your computer, and record your voice while vocalizing all vowels in turn. Generate a *spectrogram* of the data (see Fig. 9.17). Use *audacity* to cut out one single vowel, and have a look at the power spectrum. Can you format it such that you can see the first few harmonics of your vocal cords?

# References

Cooley, J. W., & Tukey, J. W. (1965). An algorithm for the machine calculation of complex Fourier series. *Mathematics of Computation, 19*(90), 297–301.

Fan, J., & Yao, Q. (2003). *Nonlinear Time Series: Nonparametric and Parametric Methods.* New York, NY: Springer.

Harris, C. M. (1998). The Fourier analysis of biological transients. *Journal of Neuroscience Methods, 83*(1), 15–34.

Heideman, M., Johnson, D., & Burrus, C. (1984). Gauss and the history of the fast fourier transform. *IEEE ASSP Magazine. 1558–1284*(4), 14–21, https://doi.org/10.1109/MASSP.1984.1162257.

Press, W., Teukolsky, S., Vetterling, W., & Flannery, S. (2007). *Numerical Recipes in C* (3rd ed.). Cambridge: Cambridge University Press.

Smith, III, J. O. (2007). *Mathematics of the Discrete Fourier Transform (DFT): With Audio Applications.* W3K Publishing.

Unpingco, J. (2014). *Python for Signal Processing.* Berlin: Springer.

# Chapter 10

# Solving Equations of Motion

The Fourier Transform represents time signals as a sum of sinusoids and is therefore well suited for systems with a constant frequency content. For temporally changing signals, however, such as the electrical current that builds up when a light switch gets flipped, it is less well suited. Such signals can be better represented with the *Laplace Transformation*, which contains not only sinusoids with constant amplitude but also exponentially growing and decaying signals.

Exponentially growing and decaying functions also often appear in the solution of differential equations. The Laplace Transform is therefore well suited to convert differential equations, which are often used to characterize system elements in the time domain, into the frequency domain. Since the Laplace Transform has the very convenient property of turning differential equations (in the time domain) into algebraic equations (in the frequency domain), it is immensely helpful in obtaining solutions to problems described by differential equations.

This chapter only gives a basic introduction into some principles of simulation and control of systems. For a more in-depth description of *control systems* and of the underlying principles and techniques, see for example Aström and Murray (2016).

## 10.1 Transfer Functions

### 10.1.1 Responses to Sinusoidal Inputs

In a first approximation, many systems can be described by a first order differential equation. Such systems are called *Linear Time Invariant Systems (LTIs)*. (The exact definition of LTI systems has already been given in Sect. 5.2.1.) One important consequence of linearity is that a sine-input always leads to a sine-output with the same frequency, with changes only to the *amplitude* and the *phase* of the oscillation (Fig. 10.1).

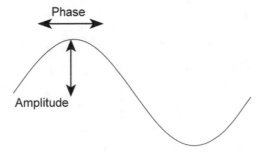

Figure 10.1: Amplitude and Phase of a sine-wave.

T. Haslwanter, *An Introduction to Hands-on Signal Analysis with Python*,
https://doi.org/10.1007/978-3-030-57903-6_10

Using Euler's formula (Eq. 1.5), sinusoidal oscillations can be expressed with $e^{j\omega t}$. Note that the only purpose of using complex numbers is to keep the mathematics as simple as possible. If the input of an LTI system is $x(t) = e^{j\omega t}$, the output must have the form

$$y(t) = r \cdot e^{j\delta} \cdot e^{j\omega t} = G(j\omega) \cdot e^{j\omega t} \tag{10.1}$$

The *gain* $r$ quantifies the change in amplitude, and $\delta$ the *phase shift* introduced by the system. So one complex number, $G(j\omega)$, completely characterizes the effect of the system on a sine input with that frequency. $G(j\omega)$ is therefore called the "*transfer function*" of the system (see also Fig. 5.2).

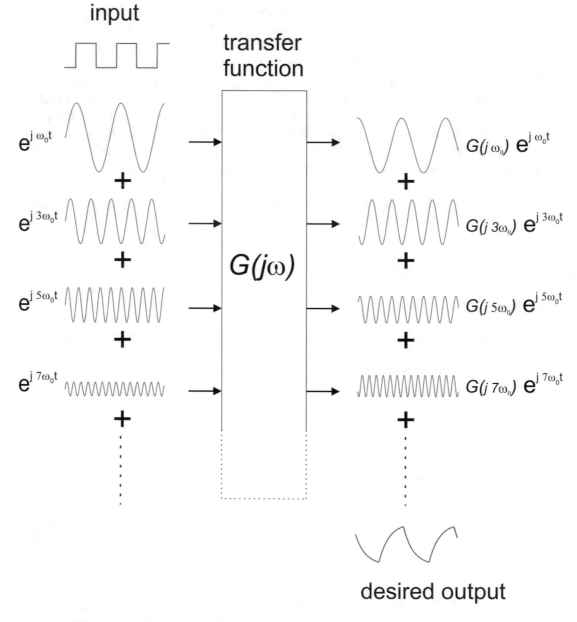

Figure 10.2: The basic idea underlying *linear systems analysis*: The *transfer function* defines for each sinusoidal input with frequency $\omega$ the change in amplitude and in phase. This change can be describes with a single complex number $G(j\omega)$.

### 10.1.2 Superposition

All linear systems obey superposition. (See also Sect. 5.2.1, where linearity and time invariance are discussed for discrete systems.) A linear system is one described by linear equations. $y = kx$ is linear ($k$ is a constant). $y = sin(x)$, $y = x^2$, $y = log(x)$ are obviously not. Even $y = x + k$ is not linear, one consequence of superposition being that double the input ($x$) should lead to double the output ($y$), and with $y = x + k$ this is not the case. The differential equation

$$a\frac{d^2x}{dt^2} + b\frac{dx}{dt} + cx = y$$

is linear.

$$a\frac{d^2x}{dt^2} + b\frac{dx}{dt}x + cx^2 = y$$

is not, for two reasons which I hope are obvious.

If the input $x(t)$ can be decomposed into a sum of sine waves, the transfer function for a linear system can quickly give us the gain and phase for each sine wave or *harmonic*. This will give us all the output sine waves, and all we have to do is add them all up and "voila!" we have the desired output.

This is illustrated in Fig. 10.2. The input $x(t)$ (the square wave is only for illustration) is decomposed into the sum of a lot of harmonics on the left using the Fourier Transform to find their amplitudes. (For discrete, time limited systems, the resulting frequencies are multiples of $\omega_0$.) Each is passed through $G(j\omega)$. $G(j\omega)$ has a gain and a phase shift which depends on the frequency. The resulting sinusoids $G(jn\omega_0)e^{jn\omega_0 t}$ can then all be added up as on the right side to produce the final desired output shown at the lower right.

Figure 10.2 illustrates the basic method of all transforms including Laplace transforms so it is important to understand the concept (if not the details). In different words, $x(t)$ is taken from the time domain by the transform into the frequency domain. There, the system's transfer function operates on the frequency components to produce output components still in the frequency domain. The inverse transform assembles those components and converts the result back into the time domain, which is where we want the answer. Obviously one couldn't do this without linearity and superposition.

## 10.2 Laplace Transformation

In the simulation of mechanical systems (or differential equations in general), the conversions from the time to the frequency domain is typically performed with the *Laplace Transform*.

Given any time course x(t), its Laplace transform X(s) is

$$X(s) = \int_0^\infty x(t) \cdot e^{-st} dt \tag{10.2}$$

The *inverse Laplace transform* is given by the following complex integral, which is known by various names such as *Bromwich integral* or *Fourier-Mellin integral*:

$$x(t) = \frac{1}{2\pi j} \lim_{T\to\infty} \int_{\gamma-jT}^{\gamma+jT} e^{st} X(s) \, ds \tag{10.3}$$

where $\gamma$ is a real number so that the contour path of integration is in the region of convergence of $X(s)$. (That sounds a bit daunting; but in practice you typically don't have to worry about those details.) "$s$" is sometimes referred to as *complex frequency*.

The Laplace transform is a generalization of Fourier transform. With Fourier transformations we have dealt only with sine waves, $e^{j\omega t}$. Put another way, we have restricted $s$ to $j\omega$ so that

$e^{st}$ was restricted to $e^{j\omega t}$. But this is unnecessary, we can let $s$ enjoy being fully complex or $s = \sigma + j\omega$ . This greatly expands the kinds of functions that $e^{st}$ can represent.

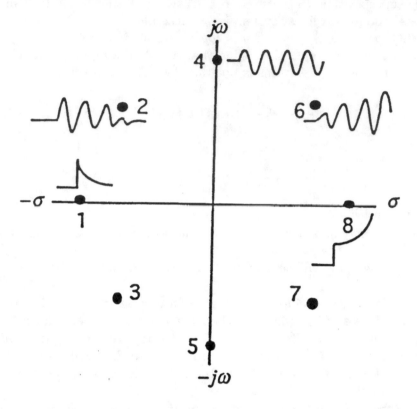

Figure 10.3: The Laplace Transform uses exponentially changing sinusoids. (From Robinson (1994)).

Figure 10.3 is a view of the s-plane with its real axis ($\sigma$) and imaginary axis ($j\omega$). It shows the variety of waveforms represented by $e^{st}$. At point 1, $\omega = 0$ and $-\sigma$ is negative so $e^{st} = e^{-\sigma t}$, which is a simple decaying exponential as shown. At points 2 and 3 (we must always consider pairs of complex points, recall from Eq. 9.3 that it took an $e^{j\omega t}$ and an $e^{-j\omega t}$ to get a real $sin(\omega t)$ or $cos(\omega t)$ ) we have $-\sigma < 0$ and $\omega \neq 0$ , so $e^{-\sigma t}e^{j\omega t}$ is a damped sine wave as shown. At points 4 and 5, $\sigma = 0$ so we are back to simple sine waves. At points 6 and 7, $\sigma > 0$ so the exponential is a rising oscillation. At $8, \sigma > 0, \omega = 0$ so we have a plain rising exponential.

So Eq. 10.3 says that $x(t)$ is made up by summing an infinite number of infinitesimal wavelets of the forms shown in Fig. 10.3, and $X(s)$ tells you how much of each wavelet is needed at each point on the s-plane. That weighting factor is given by the transform (Eq. 10.2). In terms of Fig. 10.2, x(t) is decomposed into an infinite number of wavelets as shown in Fig. 10.3, each weighted by the complex number X(s). They are then passed through the transfer function G which now is no longer $G(j\omega)$ (defined only for sine waves) but G(s) defined for the whole complex plane. The result of $X(s)G(s)$ tells you the amount of $e^{st}$ at each point on the s-plane contained in the output. Using (10.3) on $X(s)G(s)$ takes us back to the time domain and gives the output.

A very important aspect of the Laplace transform is

$$\frac{dx(t)}{dt} \xrightarrow{LaplaceTransform} s \cdot X(s) - x(0) \tag{10.4}$$

In mathematical notation:

$$\mathcal{L}[\frac{dx(t)}{dt}] = s\mathcal{L}[x(t)] - x(0)$$

where $\mathcal{L}$ indicates the Laplace transform. This equation states that a Laplace transform converts a differential equation (as a function of time) into an algebraic equation (as a function of frequency), thereby allowing us to easily solve the differential equation in the frequency domain.

## 10.2.1 Parallel Systems

As an example, consider the force developed by a muscle as a function of its innervation. The input of such a system is the innervation of the muscle by the corresponding motor nerve, the output is the muscle force. Such a system can to a first approximation be represented as a *visco-elastic* element. In these systems one is concerned with force, displacement and its rate of change, velocity.

To simulate such a system, let us start with the elastic element (Fig. 10.4).

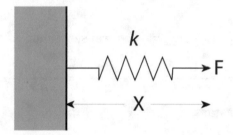

Figure 10.4: F is the force, x the length, and k the spring constant.

*Hook's Law* states

$$f(t) = k * x(t) \tag{10.5}$$

where $f$ is the force and $k$ the spring constant.

The damping by the muscle viscosity depends (similar to the shock absorbers in a car's suspension system, or the friction in a hypodermic syringe) on the rate of change of the length of the muscle. Such an element is sometimes called a "dash pot" (Fig. 10.5).

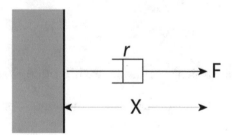

Figure 10.5: Plunger in a cylinder, with a viscosity $r$.

The relationship for the dash pot is

$$f(t) = r * \frac{dx}{dt} \xrightarrow{\mathcal{L}} r * s * X(s) \tag{10.6}$$

where $X(s)$ indicates the Laplace-transform of $x(t)$. That is, a constant force causes the element to change its length at a constant velocity. The friction coefficient $r$ characterizes the viscosity.

Put together, this combination is sometimes referred to as *Voigt-Element*, *low pass filter*, or *first order lag*, and is a good first approximation for a muscle model (Fig. 10.6). The force $F$ acting on the mass is divided between the two elements, and—since the two elements are arranged in parallel—the overall force is $F = F_k + F_r$, or

$$F(s) = (k + r * s) X(s) \tag{10.7}$$

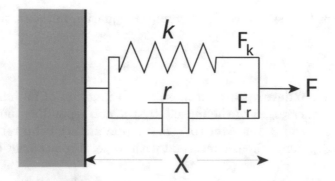

Figure 10.6: Parallel combination of a spring and a damper.

If we take F to be the input and X the output, the transfer function $G$ is

$$\frac{X(s)}{F(s)} = G(s) = \frac{1}{s*r+k} = \frac{1/k}{s*r/k+1} = \frac{1/k}{s*\tau+1} \tag{10.8}$$

where $\tau = r/k$ is the system time constant.

**Damped Oscillator**

The next level of realism (or complexity) is the addition of a *mass m* (Fig. 10.7). Compared to the Voigt element, we now also have to include the inertial force $f = m\frac{d^2x}{dt^2}$

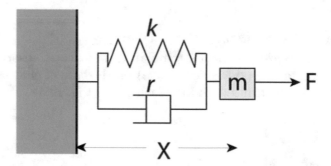

Figure 10.7: A damped mechanical oscillator, with a force F acting on it.

In that case the differential equation that describes the movement of the mass is given by

$$m*\frac{d^2x}{dt^2} + r*\frac{dx}{dt} + k*x = f(t) \tag{10.9}$$

Applying the Laplace transform gives the algebraic equation

$$m*s^2*X(s) + r*s*X(s) + k*X(s) = F(s). \tag{10.10}$$

Writing output over input, we get the transfer function for a damped oscillator

$$\frac{X}{F} = \frac{1}{m*s^2 + r*s + k} \tag{10.11}$$

## 10.2.2  Serial Systems

If two transfer functions $G(s)$ and $H(s)$ are in cascade, the combined transfer function is just their multiplication

$$Y = H*G*X. \tag{10.12}$$

Since we are only dealing with linear systems the sequence can be inverted: $G * H = H * G$.

For graphical representations such as the Bode plot below it can be convenient to plot the logarithm of the transfer gain, since the log of the combined function is just the sum of the logs (Fig. 10.8):

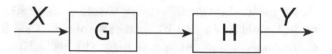

Figure 10.8: Serial combination of two transfer functions G and H.

$$\log(G * H) = \log(G) + \log(H) \tag{10.13}$$

### 10.2.3 Feedback Systems

By describing the frequency properties of individual elements the Laplace transform is a powerful tool in the simulation of complex systems. Such simulations typically start by specifying the components involved as well as their interconnections.

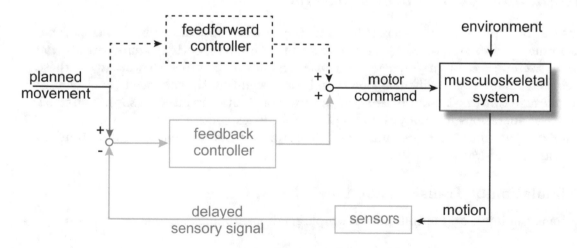

Figure 10.9: **System describing human motor control** The dotted lines indicate the feedforward control pathway, and the gray lines the feedback control pathway. Elements that form part of both are shown in solid black lines. The "musculoskeletal system" that was modeled above as damped oscillator is indicated with a drop-shadow.

For example, for human motor control the basic elements can be summarized with the scheme in Fig. 10.9. If the movement represents for example an eye movement, the desired target location is selected in the cortex (*planned movement*), and the movement started (*feedforward controller*) by activation of the oculomotor neurons in the brainstem. From there the extraocular muscles (*musculoskeletal system*) are activated, moving the eyes (*motion*) towards the target. The visual system (*sensors*) provides feedback (*feedback controller*) if the target has been reached, or if the feed forward control of the muscles must be adapted (*sensory signal*).

The individual elements of this control loop can then be simulated individually, and often linear time invariant systems form a good first approximation. For example, the muscles in this control loop can be modeled to a first approximation as the visco-elastic element described above.

To show how Python can be used to simulate such systems, we show the Python implementation of two simple examples. The top panel in Fig. 10.10 is an example of a feed forward system, with the corresponding Python implementation presented below in Sect. 10.3.1. The bottom panel shows a negative feedback system, with the Python implementation in Sect. 10.3.2.

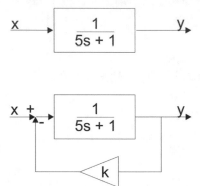

Figure 10.10: **Top: Feedforward Control** A simple first order lag, with a time constant of 5 sec. *s* is the complex frequency. **Bottom: Feedback Control** Feedback control changes the static and dynamic properties of control systems. Here a negative feedback with gain *k*.

## 10.3   Implementation of Simulations

LTI systems can be represented in different formats. Here we are only discussing *transfer functions*. This format is also supported by `scipy.signal`. However, such systems can also be represented in *state space*, or in *frequency response data* form. In order to work with these views, for conversions between different views, and for a significantly enhanced functionality the Python package `control` can be used (https://python-control.readthedocs.io). `control` implements basic operations for analysis and design of feedback control systems.

For details on the different representations of control systems and on control systems in general see Aström and Murray (2016).

### 10.3.1   Simulation of Transfer Functions

For LTI systems the input and output have a simple relationship in the frequency domain:

$$Y(s) = G(s) * X(s)$$

where the transfer function G(s) can be expressed by an algebraic function

$$G(s) = \frac{num(s)}{den(s)} = \frac{n_0 * s^0 + n_1 * s^1 + n_2 * s^2 + ...}{d_0 * s^0 + d_1 * s^1 + d_2 * s^2 + ...}$$

In other words, specifying **n** and **d**, the vectors containing the coefficients of the numerator and denominator, uniquely characterizes the transfer function. This notation can be used by computational tools to simulate the response of such a system to a given input.

For example, the response of a low-pass filter with a time-constant of 5 s (Fig. 10.10, top) has the following transfer function

$$G(s) = \frac{1}{5 * s + 1}$$

and the response can be simulated with `scipy.signal`:

```
from scipy import signal

# Define the transfer function
num = [1]
den = [tau, 1]
my_system = signal.lti(num, den)

# Simulate the feed forward response
in_signal = ...
t_out, out_signal, x_out = signal.lsim(my_system, in_signal, time)
```

🐍 python **Code:** `F12_19_feedback.py` contains the full code for the simulation of the response of a feed-forward and feedback system (Fig. 10.10) to a force step. The results are shown in Fig. 10.11.

### 10.3.2   Simulation of Feedback

The implementation of feedback can be implemented with the package `control`. Care should be taken, however, since the syntax is somewhat different from the one used in `scipy.signal`.

The code sample below shows how `control` can be used to implement the negative feedback loop shown in the bottom panel of Fig. 10.10.

```
import control

# First, define the feed forward transfer function
sys = control.TransferFunction(num, den)

# Simulate the response of the feed forward system
t_ff, out_ff = control.forced_response(sys, T=time, U=in_signal)


# Then define a feedback-loop, with gain fb_gain
sys_total = control.feedback(sys, fb_gain)
print(sys_total)     # 1 / (tau*s + (1+fb_gain))

# Simulate the response of the feedback system
t_fb, out_fb = control.forced_response(sys_total,
T=time, U=in_signal)
```

## 10.4   Bode Diagram

When we plot the transfer function of a first order lag with a time-constant of 5 sec in order to describe the behavior of the system, we get (Fig. 10.12).

In the 1930s, Hendrik Wade Bode decided to plot the gain on a log-log plot. Doing this the low frequency part of the $\omega$ axis gets stretched out, providing a clearer view of the system's frequency behavior (Fig. 10.13). In Python that can be done with

```
import matplotlib.pyplot as plt
from scipy import signal

tau = 5    # [s]
sys = signal.lti([1], [tau, 1])
```

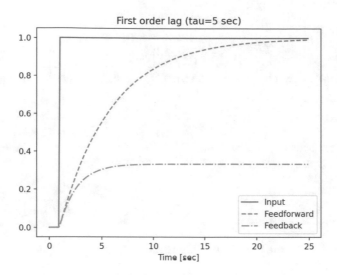

Figure 10.11: Results of the simulation of a the feedforward system shown in Fig. 10.10 top, and the corresponding system with a simple negative feedback with gain $k=2$ (Fig. 10.10 bottom). The effect of the negative feedback is to reduce the dynamic response time of the system (i.e. the system reacts faster), but it also reduces the static gain.

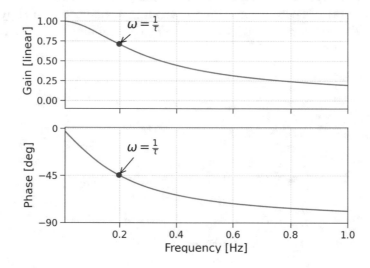

Figure 10.12: As frequency $\omega = 2\pi f$ goes up, the gain goes down, approaching zero, and the phase lag increases to 90 deg. (From F10_bode.py).

```
w, mag, phase = signal.bode(sys)

fig, axs = plt.subplots(2, 1, sharex=True)
axs[0].semilogx(w, mag)      # Bode magnitude plot
axs[0].set_ylabel('Gain [dB]')

axs[1].semilogx(w, phase)   # Bode phase plot
axs[1].set_ylabel('Phase [deg]')
axs[1].set_xlabel('Freq [rad/s]')
axs[1].set_yticks([-90, -45, 0])
plt.grid(True)

plt.show()
```

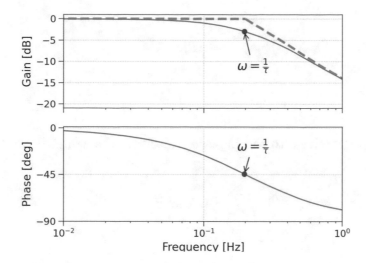

Figure 10.13: **Bode Plot** Note that in the Gain plot (top) the exact transfer function is well approximated by two straight lines which intersect at the frequency $\omega = 1/\tau$. (From F10_bode.py).

An interesting frequency for a first order lag is $\omega = 1/\tau$ , since then the gain is $|G| = 1/\sqrt{2} = 0.707$ and the phase $\angle G = -\tan^{-1}(1) = -45°$ . Below this frequency, $log(|G|)$ can be closely approximated by the horizontal dotted line at zero in Fig. 10.13 (since $log(1) = 0$ ). At frequencies above $\omega = 1/T$, $|G|$ falls off in another straight line with a slope of *20dB/dec*, which means it falls by *10* if $\omega$ increases by *10*. The phase is a linear-log plot. Thus the Bode plot in Fig. 10.13 is a very simple way to portray what a first order lag element does to sine wave inputs $e^{j\omega t}$ of any frequency.

## 10.5   Discrete Versus Continuous Systems

This section contains a summary of the most important aspects of discrete linear systems.

### 10.5.1   Laplace Transform Versus Fourier Transform

As mentioned in Eq. 10.2, the Laplace transform of a signal $f(t)$ is defined as

$$X(s) = \int_0^\infty x(t) \cdot e^{-st} dt \qquad (10.14)$$

where $t$ is real and $s = \sigma + j\omega$ is complex.

When evaluated along $\sigma = 0$, i.e. along the $s = j\omega$ axis, the Laplace transform reduces to

$$X(j\omega) = \int_0^\infty x(t) e^{-j\omega t} dt.$$

Apart from the lower transform limit, this corresponds to the Fourier transform (see Eq. 9.1). The Fourier transform is normally defined from $-\infty$, but for causal signals (i.e. $x(t) = 0$ for $t \leq 0$), there is no difference. In other words, the Fourier transform is obtained by evaluating the Laplace transform along the complex axis $(j\omega)$ in the s-plane.

### 10.5.2   Z-Transformation

While the Laplace transform is used for continuous system, the *z-transform* is used for discrete systems:

$$X_d(z) = \sum_{n=0}^{\infty} x_d(nT)z^{-n}. \tag{10.15}$$

Defining

$$z = e^{sT}, \tag{10.16}$$

the $z$ transform becomes proportional to the Laplace transform of a sampled continuous-time signal:

$$X_d(e^{sT}) = \sum_{n=0}^{\infty} x_d(nT)e^{snT}.$$

**Shift Theorem**

A delay of $\Delta$ samples in the time domain corresponds to a multiplication by $z^{-\Delta}$ in the frequency domain:

$$x(n - \Delta) \leftrightarrow z^{-\Delta}X(z), \Delta \geq 0. \tag{10.17}$$

This explains the use of $z^{-1}$ in the representations of FIR- and IIR-filters in Figs. 5.5 and 5.7.

**Convolution Theorem**

For any two signals $x$ and $y$, convolution in the time domain corresponds to the multiplication in the $z$ domain:

$$x(t_i) * y(t_i) \leftrightarrow X(z) \cdot Y(z) \tag{10.18}$$

### 10.5.3   Transfer Function and Impulse Response

Denoting the impulse response of a discrete system with $h(n)$, the transfer function $H(z)$ that defines linear, time-invariant systems is equal to the z transform of the impulse response $h(n)$:

$$H(z) = \frac{Y(z)}{X(z)} \tag{10.19}$$

**Z Transform of Difference Equations**

As stated in Eq. 5.12, the general difference equation for IIR-filters is

$$\sum_{j=0}^{m} a_j * y(n - j) = \sum_{i=0}^{k} b_i * x(n - i)$$

Taking the $z$ transform of both sides, and making use of the shift theorem (see above) leads to

$$H(z) = \frac{Y(z)}{X(z)} = \frac{b_0 + b_1 z^{-1} + \ldots + b_M z^{-M}}{1 + a_1 z^{-1} + \ldots + a_N z^{-N}} \equiv \frac{B(z)}{A(z)}$$

## 10.6  Exercises

1. **Mechanical Systems** The force response of a human muscle (as a function of the innervation) can be modeled with a "Voigt-Element", a spring and a damper arranged in parallel (top panel).

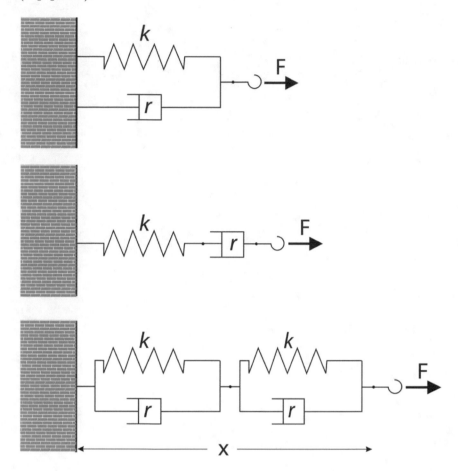

- What it the transfer function when spring and damper are arranged in parallel (Voigt-element, top)?
- What it the transfer function when spring and damper are arranged serially (middle)?
- What it the transfer function when two Voigt elements are connected in series (bottom)?
- With values of $k = 1\,\mathrm{N\,m^{-1}}$ and $r = 2\,\mathrm{N\,s\,m^{-1}}$, simulate the responses of the three configurations to a sudden force-step of 5 N.

## References

Aström, K. J., & Murray, R. M. (2016). *Feedback Systems: An Introduction for Scientists and Engineers* (2nd ed.). Princeton: Princeton University Press.

Robinson, D. (1994). *Feedback Control Systems for Biomedical Engineering.* Lecture Notes. Baltimore: Johns Hopkins University.

# Chapter 11

# Machine Learning

While this book is intended as an introduction to signal analysis, it would not be complete without demonstrating the powers of *machine learning (ML)* in Python in this final chapter. Signal analysis with ML is substantially different from the approaches presented in the previous chapters. With ML an analysis program is first presented with data, which may have been labeled, and tries to find patterns to those data. This is called "training" the ML program. This "trained" program is then used to predicts class-membership or values for new data ("prediction"). A number of packages facilitate ML in Python, such as *scikit-learn*, *Theano*, *TensorFlow*, *Keras*, and *PyTorch*. In the following I will present two small examples using *scikit-learn* (https://scikit-learn.org). *scikit-learn* is one of the most popular ML libraries for classical ML algorithms. It supports most of the supervised and unsupervised learning algorithms. It can also be used for data-mining and data-analysis, which makes it a great tool for those starting out with ML.

While ML offers tremendous powers to the skilled user, mastering it should not be underestimated. ML requires substantial knowledge and experience to be applied correctly. An excellent introduction to the area is provided by Müller and Guido (2016), and advanced information can be found in Géron (2017).

Here I want to present two examples of *supervised machine learning*. In supervised learning the program is first presented with a number of well characterized data points. From these data it tries to find rules characterizing the data. For *classification* the data points contain information about the properties of each item, as well as the corresponding *category*. Once the program has been trained, it can be applied to unknown items and determine which category these new items belong to. In *regression* tasks the program is presented with known input values and corresponding output values. From these it tries to extract the relationship between input and output data, which can then be applied to new input values.

## 11.1 Example 1: Predicting a Class

One group of ML applications deals with assigning objects to one of a group of classes. An example would be a spam-filter for email, which classifies incoming mail as either "OK" or "Spam". Such task are called "classification tasks". Another example would be the classification of flowers, based on the measured properties of the plants. Once the algorithm has been trained with known samples, the measurements of a new plant can be used to classify that plant. Or yet another example would be movement recordings: first multiple measurements of labeled movement recordings (e.g. sitting, walking, and running movements) are presented to the computer program to train an algorithm. Afterwards this algorithm can be used to classify new, unlabeled recordings (Fig. 11.1).

The example in Listing 11.1 shows a common textbook example, the prediction of a flower species based on the measured properties of the flower petals.

T. Haslwanter, *An Introduction to Hands-on Signal Analysis with Python*, https://doi.org/10.1007/978-3-030-57903-6_11

**Listing 11.1: classification.py**

```python
""" Simple classification model for "iris" data-set """

# Import the standard packages ...
import numpy as np
import matplotlib.pyplot as plt
import pandas as pd

# ... and the data and functions needed from scikit-learn
from sklearn.datasets import load_iris
from sklearn.model_selection import train_test_split
from sklearn.neighbors import KNeighborsClassifier

# Load the data
iris_dataset = load_iris()

# Show misc data information
print(f'Keys of iris_dataset: {iris_dataset.keys()}')
print(iris_dataset['DESCR'])
print(f"Target names: {iris_dataset['target_names']}")
print(f"Feature names: {iris_dataset['feature_names']}")
print(f"Shape of data: {iris_dataset['data'].shape}")
print(f"First 5 rows of data:\n {iris_dataset['data'][:5]}")

# Split data into training- and test-set
X_train, X_test, y_train, y_test = train_test_split(iris_dataset['data'],
        iris_dataset['target'], random_state=0)

# Define and train the network
knn = KNeighborsClassifier(n_neighbors=3)
knn.fit(X_train, y_train)

# Show how accurate it is
print(f'Test set score: {knn.score(X_test, y_test):.2f}')

# Plot the data ---------------------------
# First bring the data into a pandas DataFrame, and group them by "species"
plot_data = np.column_stack( (X_train, y_train) )
df = pd.DataFrame(data=plot_data,
                  columns=['Prop_1', 'Prop_2', 'Prop_3', 'Prop_4', 'species'])
groups = df.groupby('species')

# Plot the groups
fig, axs = plt.subplots(1,2)
for name, group in groups:
    axs[0].plot(group.Prop_1, group.Prop_2, 'o')
    axs[1].plot(group.Prop_3, group.Prop_4, 'o')

axs[0].set_xlabel('Property 1')
axs[0].set_ylabel('Property 2')
axs[1].set_xlabel('Property 3')
axs[1].set_ylabel('Property 4')

# Take an arbitrary new sample, and plot it
new_sample = np.array([[7, 3.5, 6, 2]])
axs[0].plot(*new_sample[0,:2], 'r+', ms=18)
axs[1].plot(*new_sample[0,-2:], 'r+', ms=18)

axs[1].legend( list(iris_dataset['target_names']) + ['predicted'],
               loc='upper left')

# Classify it, and show the result
```

```
61 classified = knn.predict(new_sample)
62 plt.text(3, 0.1, f"Predicted: {iris_dataset['target_names'][classified]}")
63
64 plt.tight_layout()
65
66 # To save to an out-file with my default formatting
67 out_file = 'ml_classified.jpg'
68 plt.savefig(out_file, dpi=200, quality=90)
69 print(f'Image saved to {out_file}')
70
71 plt.show()
```

**Lines 3-22** Import the required packages, load the data, and display some data characteristics.

**Line 25** Here the interesting stuff starts. We want to do two things with the data available: i) train our model, and ii) test how well it performs (otherwise, we would have no idea if our approach works or fails!). Since we have to use independent samples for those two tasks, the command `train_test_split` is used to randomly split the available data into the two corresponding groups.

**Line 28** `scikit-learn` includes a range of models for the classification of data: *support vector machines*, *random forest models*, *nearest neighbor models* etc. Here we select the `KNeighborsClassifier`, and specify that the number of neighbors to be taken into consideration is `n_neighbors=3`. This means that when a new point has to be classified, the 3 nearest neighbors from the training data set are selected, and the new point is *predicted* to have the same type as the majority of those neighbors.

**Line 29** This one line is all that it takes to train the model!!

**Line 32** In order to test the accuracy of the model, all the test-data are classified and the predicted classes compared to the actual classes. The resulting `score` is displayed.

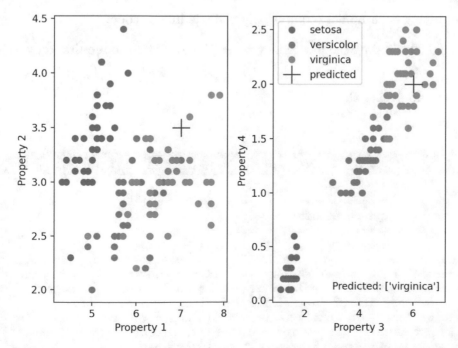

Figure 11.1: Properties of the training set (colored points), and classification of a new sample (red cross) (from Listing 11.1)

The beauty of `scikit-learn` is that it uses a very consistent naming scheme. So if we want to test a different ML model, we only have to change one line of our code, i.e. line 28! This makes it incredibly simple to evaluate and compare different models. Thereby the options defining the model make it possible to fine-tune the model to the task at hand.

## 11.2   Example 2: Predicting a Value

Another group of ML task deals with the quantitative prediction of relationships between different parameters. These tasks are called "regression tasks". The input parameters in regression tasks are sometimes called *exogenous variables* or *regressors*. Based on these, regressions try to determine the most likely corresponding output variable, sometimes called *predicted variable*, *endogenous variable*, or *regressand*. Thereby we may make assumptions about the shape of our trajectory, for example we may assume that the relationship is linear. In the example presented below we do *not* restrict our fit to a polynomial, or to a known function like an exponential decay, but simply assume that there is a smooth relationship between our regressor and the predicted variable (Fig. 11.2). To illustrate how easy it is to compare different models, we compare a fit with a technique called "support vector regression" to a fit with a model based on "kernel ridge regression". (Don't worry if you do not understand *how* these models work here, since we only want to demonstrate how easy it is to perform such tasks with *scikit-learn*. Details on the underlying algorithms can be found for example in Müller and Guido (2016) and in Géron (2017).)

And since we do not know what the best parameters are for each of those two types of model, we go one step further and also make a parameter search over a specified grid of parameter values with the command `GridSearchCV`:

**Lines 1-39** Header, and import of the required packages

**Lines 41-50** Generation of a noisy dummy curve, which we want to fit with a smooth model

**Lines 54-60** Here we define that we want to test models with *support vector regression (SVR)* and *KernelRidge*-models, and specify for each the parameter grid that we want to sweep.

**Lines 62-63** Again, fitting each model requires only a single line of code!

**Lines 65-66** And similarly, finding the most likely value for new data also requires only a single line of code.

**Lines 68-88** Plotting and saving the results.

Listing 11.2: regression.py

```
1  """
2  =============================================
3  Comparison of kernel ridge regression and SVR
4  =============================================
5
6  Both kernel ridge regression (KRR) and SVR learn a non-linear function by
7  employing the kernel trick, i.e., they learn a linear function in the space
8  induced by the respective kernel which corresponds to a non-linear function in
9  the original space. They differ in the loss functions (ridge versus
10 epsilon-insensitive loss). In contrast to SVR, fitting a KRR can be done in
11 closed-form and is typically faster for medium-sized datasets. On the other
12 hand, the learned model is non-sparse and thus slower than SVR at
13 prediction-time.
14
```

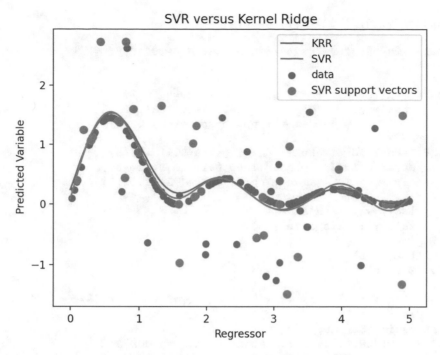

Figure 11.2: Fitting a smooth curve to an unknown set of noisy data, with two different types of models: support vector regression, and kernel ridge regression. (From Listing 11.2)

```
15 This example illustrates both methods on an artificial dataset, which
16 consists of a sinusoidal target function and strong noise added to every fifth
17 datapoint. The first figure compares the learned model of KRR and SVR when both
18 complexity/regularization and bandwidth of the RBF kernel are optimized using
19 grid-search. The learned functions are very similar; however, fitting KRR is
20 approx. seven times faster than fitting SVR (both with grid-search). However,
21 prediction of 100000 target values is more than tree times faster with SVR
22 since it has learned a sparse model using only approx. 1/3 of the 100 training
23 datapoints as support vectors.
24
25 """
26
27 # Authors: original code by Jan Hendrik Metzen <jhm@informatik.uni-bremen.de>;
28 #          code elements extracted by Thomas Haslwanter
29 # License: BSD 3 clause
30
31 # Import the required packages
32 import numpy as np
33 import time
34
35 from sklearn.svm import SVR
36 from sklearn.model_selection import GridSearchCV
37 from sklearn.kernel_ridge import KernelRidge
38 import matplotlib.pyplot as plt
39
40 # #######################################################################
41 # Generate sample data
42 rng = np.random.RandomState(0)
43 X = 5 * rng.rand(10000, 1)
44 y = (np.sin(2*X)**2/X).ravel()
45
46 # Add noise to targets
47 y[::5] += 3 * (0.5 - rng.rand(X.shape[0] // 5))
48
```

```
49  X_plot = np.linspace(0, 5, 100000)[:, None]
50
51  # ############################################################################
52  # Fit regression model
53  train_size = 100
54  svr = GridSearchCV(SVR(kernel='rbf', gamma=0.1), cv=5,
55                     param_grid={"C": [1e0, 1e1, 1e2, 1e3],
56                                 "gamma": np.logspace(-2, 2, 5)})
57
58  kr = GridSearchCV(KernelRidge(kernel='rbf', gamma=0.1), cv=5,
59                    param_grid={"alpha": [1e0, 0.1, 1e-2, 1e-3],
60                                "gamma": np.logspace(-2, 2, 5)})
61
62  svr.fit(X[:train_size], y[:train_size])
63  kr.fit(X[:train_size], y[:train_size])
64
65  y_svr = svr.predict(X_plot)
66  y_kr = kr.predict(X_plot)
67
68  # ############################################################################
69  # Look at the results
70  sv_ind = svr.best_estimator_.support_
71  plt.scatter(X[:200], y[:200], c='C0', label='data', zorder=1)
72  plt.plot(X_plot, y_kr, c='C0', label='KRR')
73
74  plt.scatter(X[sv_ind], y[sv_ind], c='C1', s=50, label='SVR support vectors',
75              zorder=2 )
76  plt.plot(X_plot, y_svr, c='C1', label='SVR ')
77
78  plt.xlabel('Regressor')
79  plt.ylabel('Predicted Variable')
80  plt.title('SVR versus Kernel Ridge')
81  plt.legend()
82
83  # To save to an out-file with my default formatting
84  out_file = 'fit_regression.jpg'
85  plt.savefig(out_file, dpi=200, quality=90)
86  print(f'Image saved to {out_file}')
87
88  plt.show()
```

# References

Géron, A. (2017). *Hands-On Machine Learning with Scikit-Learn and TensorFlow*. O'Reilley.

Müller, A., & Guido, S. (2016). *Introduction to Machine Learning with Python*. O'Reilley.

# Chapter 12

# Useful Programming Tools

To round off the book, this chapter contains information on how to refine your programming skills. It will help you to get most value out of the time you invest in programming, and ensures that your programs can be re-used and adapted later on.

## 12.1  Debugger

A *Debugger* is a tool that lets you interrupt the program execution at chosen locations in your program, or if an error occurs. This can be tremendously helpful for finding errors and solving problems. On the downside, additional files need to be loaded in order to *run the debugger*, vs. simply *executing the code*. Take care that you distinguish between these two modes, since the debugging mode can be significantly slower.

> **Tip:** Take the time to get to know the debugger in the IDE that you are using. This will save you a lot of time later on when developing programs.

## 12.2  Code Versioning with *git*

### 12.2.1  Overview

Computer programs rarely come out perfect on first try. Typically they are developed iteratively, by successively eliminating known errors. *Version control* programs (such as *git*), also known as *revision control* programs, allow tracking only the modifications, and storing previous versions of the program under development. If the latest changes then cause a new problem, it is easy to compare them to earlier versions and to restore the program to a previous state.

If you are developing computer software, I strongly recommend the use of *git*. It can be used locally, with very little overhead. And it can also be used to maintain and manage a remote backup copy of the programs. While the real power of *git* lies in its features for collaboration, I have been very happy with it for my own data and software. An introduction to *git* goes beyond the scope of this book, but a very good instruction is available under https://git-scm.com/. Good, short and simple starting instructions—in many languages—can be found at http://rogerdudler.github.io/git-guide/.

A source of confusion can be the difference between "git" and "github":

**git** is a version control program, and is well integrated in most Python IDEs.

**github** is a website (https://github.com/) frequently used to share code, and is now owned by Microsoft. It is the place where the source code for the majority of open source Python packages is hosted. While one can also download source code from there, it is more efficient to use *git* for this task.

T. Haslwanter, *An Introduction to Hands-on Signal Analysis with Python*,
https://doi.org/10.1007/978-3-030-57903-6_12

Under Windows *tortoisegit* (https://tortoisegit.org/) provides a very useful Windows shell interface for *git*. (Note that *git* and *tortoisegit* have to be installed separately!) For other operating systems, and if more functionality is required under Windows, *smartgit* is a good choice (https://www.syntevo.com/smartgit/). While being a commercial product, it is free for academic and non-commercial use.

### 12.2.2  Installation and Interfaces

*git* can be downloaded for free from https://git-scm.com/. There one also finds very good documentation, and also help getting started with *git*. Once it is installed it can be run in different ways:

- It can be run from a graphical user interface (GUI). Numerous GUIs exist (see https://git-scm.com/downloads/guis). Note that *git* has to be installed separate from the GUI interfaces.

- It can be run from the command-line. If you use the standard command-line tool, you first have to ensure that git.exe is part of the system path.

- *git for windows* https://gitforwindows.org/ also comes with a *git Bash*, which provides a Bash emulation used to run *git* from the command line.[1] Unix and Linux users should feel right at home, as the Bash emulation behaves just like the git command in Linux and Unix environments.

### 12.2.3  Examples

**TortoiseGit**

In order to clone a repository (e.g. https://github.com/thomas-haslwanter/sapy.git) in *tortoisegit* from *github* to your computer, you simply have to right-click on the folder where you want the repository to be installed, select Git Clone..., and enter the repository name—and the whole repository will be cloned there. Done!

**Command-line**

You can (i) start a new git repository, (ii) add the file test.txt to this repository, and (iii) commit this file with the following command sequence:

```
git init
git add test.txt
git commit -a -m "This is the first commit"
```

The options for git commit specify "to commit all staged files" (-a), with the message "This is the first commit" (-m ''This is the first commit''). 
Cloning an existing repository is even simpler: to obtain for example a copy of the repository that goes with this book, simply go to the directory where you want to have it and type

```
git clone https://github.com/thomas-haslwanter/sapy.git
```

## 12.3  Test Tools

Testing is arguably the most underestimated aspect of professional coding. But testing is indispensable for the generation of trustworthy programs. Different frameworks can be used for test-

---

[1](*Bash* is a Unix shell and command language.).

ing source code. A commonly used framework are *function-based unit tests*, or short "unittests". Function-based tests subscribe to the *xUnit* testing philosophy. One advantage of unittests is that they are already incorporated in the basic Python packages (https://docs.python.org/3/library/unittest.html).

Another popular framework is *nose* (https://nose.readthedocs.io/), which extends unittests.

But probably the easiest way to start with testing in Python is *pytest* (https://pytest.org). Thereby it is worth noting that *unittests* and *nose* test suites can be also run by *pytest*. The *pytest* framework makes it easy to write small tests, yet scales to support complex functional testing for applications and libraries.

To show the principle behind testing, let me give an example of a very simple test in the file `test_sample.py`:

```python
# content of test_sample.py
def inc(x):
    """The function to be tested. Increments inputs by 1."""
    return x + 1


def test_answer():
    """The test to check that the function 'inc' provides the correct result."""
    assert inc(4) == 5
```

The code containing the function and the testing can also be split into two separate files. To demonstrate the effects of a coding error I generate a second file file `functions.py`, containing

```python
def inc(x):
    """Increments inputs by 1. Is correct."""
    return x + 1

def dec(x):
    """Is supposed to decrement inputs by 1. Contains a mistake."""
    return x - 2
```

and a third file called `test_function.py`, that tests the functions in `functions.py` with

```python
import functions as fcn

def test_inc():
    """The test to check that the function 'inc' provides the correct result."""
    assert fcn.inc(4)==5

def test_dec():
    """The test to check that the function 'dec' provides the correct result."""
    assert fcn.dec(4)==3
```

With the three files (`test_sample.py`, `functions.py`, `test_functions.py`) in one folder, we can open that folder in a command-line terminal and simply type `pytest`. This results on my computer in

```
=================== test session starts ====================================
platform win32 -- Python 3.7.6, pytest-5.3.2, py-1.8.0, pluggy-0.13.1
rootdir: D:\Users\thomas\Data\CloudStation\Books\sapy\testing
plugins: hypothesis-4.53.3
collected 3 items

test_functions.py .F                                              [ 66%]
test_sample.py .                                                  [100%]

===================== FAILURES =============================================
_____ test_dec _____
```

```
def test_dec():
"""The test to check that the function 'dec' provides the correct result."""
>       assert fcn.dec(4)==3
E       assert 2 == 3
E        +  where 2 = <function dec at 0x000001A2881E04C8>(4)
E        +    where <function dec at 0x000001A2881E04C8> = fcn.dec

test_functions.py:13: AssertionError
=================== 1 failed, 2 passed in 0.06s ===============================
```

*pytest* goes through all files in the folder, tries to recognize the files to be tested by their name, and runs the corresponding tests. The number of "." in the output after the test-name indicate how many tests were run successfully in that module (one successful test in `test_functions.py`, and one in `test_sample`). And "F" indicates tests that failed (one in `test_functions.py`). The final message `1 failed, 2 passed` summarizes the test results.

## 12.4   Graphical User Interfaces (GUIs)

Many applications become a lot more accessible to the user if they provide a GUI. There are two reasons for that: (1) Many people are scared of the command-line, for the simple reason that they don't know (a) how to get help, and (b) what to type next. (2) GUIs reduce the number of choices, making it more obvious to the user what can and/or should be done next.

One can start to program a GUI by using a GUI framework, such as *tkinter*, *Wx*, or *Qt* directly. However, this approach should be restricted to programmers who already have experience in object-oriented programming and in user interface design. The second option is to use a package that simplifies the use of GUIs. While *PySimpleGUI* is a rather young package, it offers a good balance between usability and power.

The most common tasks for the user interface are:

- selecting an existing file for data input,

- selecting a new (or existing) file for data output, and

- selecting a directory.

The following section shows how to complete these tasks with *PySimpleGUI*. For other applications, check out https://github.com/PySimpleGUI/PySimpleGUI.

### 12.4.1   PySimpleGUI—Examples

**Selecting an Existing File**

This can be done with

```
import PySimpleGUI as sg

layout = [[sg.Text('Filename')],
          [sg.Input(), sg.FileBrowse()],
      [sg.OK(), sg.Cancel()]]

window = sg.Window('Get filename example', layout)

event, values = window.Read()
```

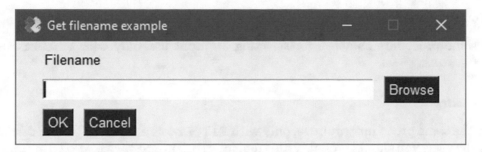

Figure 12.1: File Dialog from PySimpleGUI.

The parameter `event` is 'OK' or 'Cancel'. And `values` is a Python dictionary containing the filename selected and the filename in the text window (which is usually the same as the file selected) at the time the "OK" or "Cancel" button was pushed.

Directory, file name, and extension can be extracted with the standard Python package `pathlib`:

```
import pathlib
file_name = values[0]
path = pathlib.Path(file_name)

directory = path.parent    # e.g. WindowsPath('C:/Users/p20529/Data/Python')
full_file = path.name      # e.g. 'test.py'
extension = path.suffix    # e.g. '.py'
```

If instead of the interface in Fig. 12.1 one prefers a file-browser for file selection (see Fig. 12.2), the following command can be used:

```
import PySimpleGUI as sg
filename = sg.popup_get_file('', no_window=True)
```

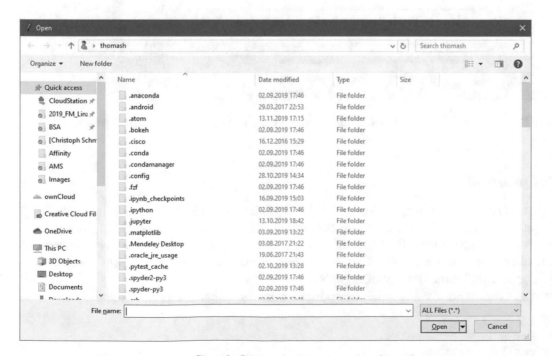

Figure 12.2: PySimpleGUI File Browser for file selection.

### Selecting an Output File

... is exactly the same procedure as above for selecting an input file, only with `FileBrowse` replaced by `FileSaveAs`.

### Selecting a Directory

... is again exactly the same procedure as above, only with `FileBrowse` replaced by `FolderBrowse` (and to be nice to the user, you should probably also replace `'Filename'` with `'Foldername'`).

### Embedding Matplotlib

Interfaces with PySimpleGUI can also incorporate Matplotlib figures (Fig. 12.3). (See also the corresponding Exercise for this chapter.)

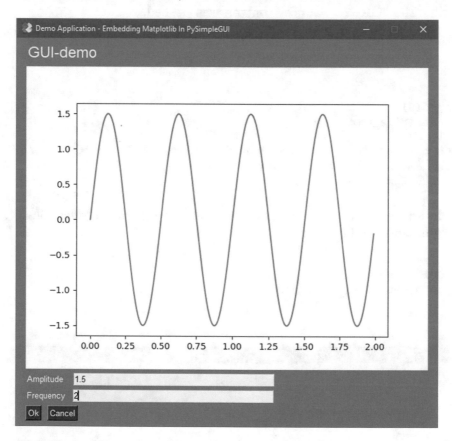

Figure 12.3: Embedding Matplotlib in PySimpleGUI. See also Exercise 2 below.

## 12.4.2  PyQtGraph

*PyQtGraph* (http://pyqtgraph.org/) is a pure Python graphics and GUI library built on *PyQt5 / PySide* and *numpy*. It is intended for use in mathematics, scientific, and engineering applications. Despite being written entirely in Python, the library is very fast due to its heavy leverage of *numpy* for number crunching and *Qt*'s *Graphics View Framework* for fast display. This makes it very useful for the real-time display of signals.

The following command gives a quick introduction to the capabilities of *PyQtGraph*:

```
from pyqtgraph import examples
examples.run()
```

For example, the different line- and scatter-plots available in *PyQtGraph* are shown in Fig. 12.4.

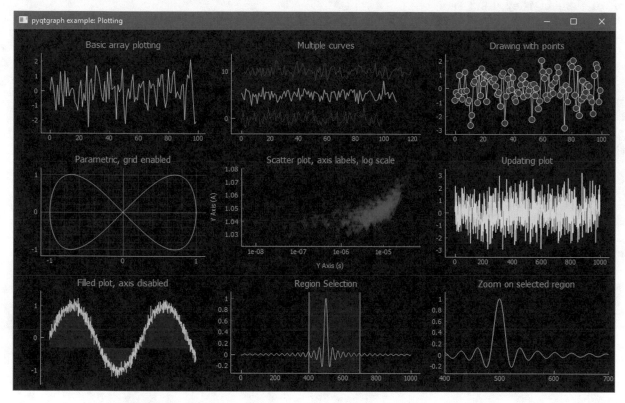

Figure 12.4: An overview of line- and scatter-plots in *PyQtGraph*.

### 12.4.3   Tips for User Interface

Development of a GUI should proceed systematically: first gather information about the user and think about the requirements; then design and implement the GUI; and afterwards, have a real user test it before the final release.

- Before you start, think about the people who will use the interface: what do they really need, what do they know, and what are the work processes in which it will be used?

  - Is this program used on a daily basis? Or is it just once-in-a while?

  - Who is using the program? Is it you (or some other expert in the field)? Or is it someone who has no experience in the application?

- Next, spend some time designing it using simple, quick tools. A pen and paper, or paper and scissors, are excellent starting points!

  - If the interface is for experts, keyboard-shortcuts should be used to accelerate frequently used tasks; if it is intended for novices, buttons are preferable.

  - If the interface is for daily use, it should be optimized for efficiency, e.g. by implementing keyboard shortcuts for frequently used tasks; if it is used just once in a while, the interface should be as self-explanatory as possible, and buttons are preferable to keyboard shortcuts.

  - When designing the GUI it is particularly important that it fits the expectations of the user, so it helps to make it similar to user interfaces that people already know.

- If the user interface isn't intended for yourself, it usually pays off to have a user quickly test it. I am amazed again and again that there may be a word that they don't understand or an element they don't see.

## 12.5   Exercises

1. **Using a Debugger** Open the code-quantlet `debug_demo.py` in an editor/debugger of your choice, and proceed to the first error. At the point of error, check where you are in the main program and where and in which function the error occurs. Check the local and global variables at the point of error, and try to execute a few python commands—using these variables—in the debugger.

2. **Construction of a GUI (hard)** create a graphical user interface (GUI, see last chapter) for this function, as shown in the Fig. 12.3.

# Appendix A

# Python Programs

All programs listed here, as well as some additional programs and the data need to run them, can be downloaded from the git repository https://github.com/thomas-haslwanter/sapy. That repository also contains Jupyter notebooks with additional topic that are relevant to signal processing, but go beyond the scope of the book (for example details of sound processing, etc.)

**Listing A.1: show_pandas.py**

```python
"""Short demonstration of data handling with Pandas

    Lines 76-85:    Show different options for handling nan's
    Lines 96-100:   Grouping, and summary information
    Line 126:       One-line boxplot of both groups
"""

# author:   Thomas Haslwanter
# date:     April-2021

# Import the required packages
import numpy as np
import matplotlib.pyplot as plt
import pandas as pd
from scipy import stats


def generate_data():
    """Generate dummy data, containing the height of 100 men and 100 women

    The return values are a Pandas DataFrame, looking as follows:
    height  gender
    185     male
    166     female
    172     female
    177     male
    etc ....

    """

    # Enter mean and standard deviation, for men and women
    height = pd.DataFrame({
        'gender':['male', 'female'],
        'mean':[176.0, 162.6],
        'std':[7.1, 7.1]
        })
```

T. Haslwanter, *An Introduction to Hands-on Signal Analysis with Python*,

```python
    # Make the "gender" the label for the row-index, and display the values
    height = height.set_index('gender')
    print('Pandas DataFrame for the height of men and women:')
    print(height)

    def make_samples(mean=170, std=10):
        """Generates 100 random samples from a normal distribution"""
        return stats.norm(mean, std).rvs(100)

    # For men and women, generate DataFrames containing height and gender
    height_dict = height.transpose().to_dict()
    print(f'Values of females only: {height_dict["female"]}')

    male   = pd.DataFrame({
        'height':make_samples(**height_dict['male']),
        'gender':'male'
        })
    female = pd.DataFrame({
        'height':make_samples(**height_dict['female']),
        'gender':'female'
        })

    # Combine the two DataFrames, mix them, and re-set the index
    data = male.append(female)
    data = data.sample(n=200)
    data = data.reset_index(drop=True)

    return data


def handle_nans():
    """Show some of the options of handling nan-s in Pandas"""

    print('--- Handling nan-s in Pandas ---')

    # Generate data containing "nan":
    x = np.arange(7, dtype=float)
    y = x**2

    x[3] = np.nan
    y[ [2,5] ] = np.nan

    # Put them in a Pandas DataFrame
    df = pd.DataFrame({'x':x, 'y':y})
    print(df)

    # Different ways of handling the "nan"s in a DataFrame:

    print('Drop all lines containint nan-s:')
    print(df.dropna())        # Drop all rows containing nan-s

    print('Replaced with the next-lower value:')
    print(df.fillna(method='pad'))   # Replace with the next-lower value

    print('Replaced with an interpolated value:')
    print(df.interpolate())          # Replace with an interpolated value


def two_categories():
    """Show how data with two categories can be handled with Pandas"""

    print('--- Grouping data in Pandas ---')
```

```python
    # Generate some dummy data, in the shape of a Pandas DataFrame
    df = generate_data()

    # Group them by gender
    grouped = df.groupby('gender')

    # Basic statistics
    print(grouped.describe())

    # On the left, show the two groups as a scatter-plot, with labels added
    fig, axs = plt.subplots(1,2)
    for name, group in grouped:
        axs[0].plot(group.height, 'o', label=name)

    axs[0].set_ylabel('Height [cm]')
    axs[0].legend()

    # If you only want the height-values of each group
    males   = grouped.get_group('male').height.values
    females = grouped.get_group('female').height.values
    df_mf = pd.DataFrame({'male': males,
                          'female': females})
    df_mf.boxplot(ax=axs[1])   # shows this plot in the right panel

    # To save to an out-file with my default formatting
    out_file = 'pandas.jpg'
    plt.savefig(out_file, dpi=200, quality=90)
    print(f'Image saved to {out_file}')

    plt.show()

    # For a standalone figure, the boxplot of the two groups can also be
    # generated with a single command:
    grouped.boxplot()
    plt.show()


if __name__ == '__main__':
    # Control the precision of pandas-output
    pd.options.display.float_format = '{:5.1f}'.format

    two_categorics()
    handle_nans()
```

Listing A.2: F5_filter_characteristics.py

```python
""" Shows how filters can be characterized. """

# author:   Thomas Haslwanter
# date:     April-2021


# Import the required packages
import numpy as np
import matplotlib.pyplot as plt
from scipy import signal
from typing import Tuple


def impulse_response(a, b, ax) -> None:
    """Show the impulse response of an IIR-filter.

    Parameters
```

```
        ----------
        a : array_like
            feedforward coefficients ('1' for FIR-filter)
        b : array_like
            feedback coefficients
        ax : mpl-axis
            plot-axis for the impulse response

        """

        # Define the impulse ...
        xImpulse = np.zeros(20)
        xImpulse[5] = 1

        # ... and find the impulse-response
        yImpulse = signal.lfilter(b, a, xImpulse)

        # Plot input and response
        ax.plot(xImpulse, '*-', label='Impulse')
        ax.plot(yImpulse, '*-', label='Response')
        ax.legend()
        ax.set_ylabel('Impulse Response')
        ax.set_xticks(np.arange(0, len(xImpulse), 5))
        ax.tick_params(axis='x', labelbottom=False)


def step_response(a, b, ax) -> None:
    """Show the impulse response of an IIR-filter.

    Parameters
    ----------
    a : array_like
        feedforward coefficients ('1' for FIR-filter)
    b : array_like
        feedback coefficients
    ax : mpl-axis
        plot-axis for the impulse response
    """

    # Define the step ...
    xStep = np.zeros(20)
    xStep[5:] = 1

    # ... and find the step-response
    yStep = signal.lfilter(b, a, xStep)

    # Plot step and response
    ax.plot(xStep, '*-', label='Step')
    ax.plot(yStep, '*-', label='Response')
    ax.legend(loc='lower right')
    ax.set_xticks( np.arange(0, len(xStep), 5) )
    ax.set_xlabel('n * T')
    ax.set_ylabel('Step Response')


def freq_response(a, b) -> Tuple[float, complex]:
    """ Show the impulse response of an IIR-filter.

    Parameters
    ----------
    a : array_like
        feedforward coefficients ('1' for FIR-filter)
    b : array_like
```

```
    feedback coefficients

Returns
-------
w : selected radial frequency of
h : complex gain for w
"""

## Frequency Response
w, h = signal.freqz(b, a, fs=2)    # Calculate the normalized values
# Plot them, in a new figure
fig, axs = plt.subplots(2, 1, sharex=True)

axs[0].plot(w, 20*np.log10( np.abs(h) ))
axs[0].set_ylim([-40, 0])
axs[0].set_ylabel('Magnitude [dB]')
axs[0].set_title('Frequency Response')

axs[1].plot(w, np.rad2deg(np.arctan2(h.imag, h.real)))
axs[1].set_ylabel('Phase [deg]')
axs[1].set_xlabel('Normalized Frequency (x pi rad/sample)')
axs[1].set_xlim([0, 1])

selFreq_val = 0.22 # Select a frequency point in the normalized response
selFreq_nr = np.argmin( np.abs(w-selFreq_val) )
selFreq_w = w[selFreq_nr]  # Value on plot

# Find gain and phase for the selected frequency
selFreq_h = h[selFreq_nr]
gain = np.abs(selFreq_h)
phase = np.rad2deg(np.angle(selFreq_h))

# Show it on the plot
dB = 20*np.log10(gain)
axs[0].plot(selFreq_w, 20*np.log10( np.abs(selFreq_h) ), 'b*')
axs[1].plot(selFreq_w,
            np.rad2deg(np.arctan2(selFreq_h.imag,selFreq_h.real)), 'b*')

plt.show()

return (selFreq_w, selFreq_h)


def show_filterEffect(w: float, h: complex) -> None:
    """ Demonstrate the filter effect on the selected frequency.

    Parameters
    ----------
    w : radial frequency
    h : complex gain
    """

    # Convert the normalized frequency to an absolute frequency
    rate = 1000

    nyq = rate/2
    dt = 1/rate
    freq = w * nyq    # Freqency in Hz, for the selected sample rate

    # Correct gain and phase
    gain = np.abs(h)
    phase = np.rad2deg(np.arctan2(h.imag, h.real))
```

```python
    # Calculate the input and output sine, for 0.04 sec
    t = np.arange(0, 0.04, dt)
    sin_in = np.sin(2*np.pi * freq * t)
    sin_out = signal.lfilter(b, a, sin_in)

    # Plot them
    plt.plot(t, sin_in, label='Input')
    plt.plot(t, sin_out, label='Output')

    plt.title(f'Input and Response for {freq:4.1f} Hz, sampled at {rate}  Hz')
    plt.xlabel('Time [s]')
    plt.ylabel('Signal')

    # Estimate gain and phase-shift from the location of the second maximum
    # First find the two maxima (input and output)
    secondCycle = np.where( (t > 1/freq) & (t < (2/freq) ) )[0]

    secondMaxIn = np.max(sin_in[secondCycle])
    indexSecondMaxIn = np.argmax(sin_in[secondCycle])
    tMaxIn = t[secondCycle[indexSecondMaxIn]]

    secondMaxFiltered = np.max(sin_out[secondCycle])
    indexSecondMaxFiltered = np.argmax(sin_out[secondCycle])
    tMaxOut = t[secondCycle[indexSecondMaxFiltered]]

    # Estimate gain and phase-shift from them
    gain_est = secondMaxFiltered / secondMaxIn
    phase_est = (tMaxIn-tMaxOut)*360*freq

    # Plot them
    plt.plot(tMaxIn, secondMaxIn, 'b*')
    plt.plot(tMaxOut, secondMaxFiltered, 'r*')
    # legend('Input', 'Response', 'maxInput', 'maxResponse')
    plt.show()

    print(f'Correct gain and phase: {gain:4.2f}, and {phase:5.1f} deg')
    print(f'Numerical estimation: {gain_est:4.2f}, and {phase_est:5.1f} deg')

    # If you want to define the figure format, add the following:
    #fig = gcf
    #fig.PaperUnits = 'inches'
    #fig.PaperPosition = [0 0 6 3]


if __name__ == '__main__':

    ## Generate coefficients for an averaging filter (FIR)
    len_filter = 5
    b = np.ones(len_filter)/len_filter
    a = 1

    fig, axs = plt.subplots(2, 1, sharex=True)

    impulse_response(a, b, axs[0])
    step_response(a, b, axs[1])
    plt.show()

    w, h = freq_response(a, b)

    show_filterEffect(w, h)
```

Listing A.3: morphology.py

```python
""" Demonstration of basic morphological operations """

# author:   Thomas Haslwanter
# date:     April-2021

# Import the required packages
import numpy as np
import matplotlib.pyplot as plt
from skimage import morphology

# Convenience functions ensuring consistent styling and folders
from utilities.my_style import set_fonts, show_data


def show_modImage(image, function: str, ax, title: str) -> None:
    """ Perform a morphological operation on an image, and display it.

    Parameters
    ----------
    image : 2D ndarray
            Image data
    function : name of function from the module skimage.morphology to be
               applied to the data
    ax : Matplotlib axis
         For the generation of the plots
    title : title for the subplot
    """

    fcn = getattr(morphology, function)
    ax.imshow(fcn(image, selem=selem))
    ax.set_title(title)
    ax.get_xaxis().set_visible(False)
    ax.get_yaxis().set_visible(False)


if __name__=='__main__':
    set_fonts(14)

    # Generate the base image
    data = np.zeros( (99,99) )
    data[34:66, 33:67] = 1
    data[85:87, 85:87] = 1
    data[49:51, 49:51] = 0

    # Show the original image
    fig, ax = plt.subplots()
    plt.gray()
    ax.imshow(data)
    ax.set_title('Original Image')
    ax.get_xaxis().set_visible(False)
    ax.get_yaxis().set_visible(False)

    out_file = 'Square.jpg'
    show_data(out_file)

    # Perform the morphological operations
    selem = morphology.square(5)
    fig, axs = plt.subplots(2,2, figsize=(8,8))

    show_modImage(data, 'binary_erosion',  axs[0][0], 'Eroded')
    show_modImage(data, 'binary_dilation', axs[0][1], 'Dilated')
    show_modImage(data, 'binary_opening',  axs[1][0], 'Opened (Dilation after
        Erosion)')
```

```
show_modImage(data, 'binary_closing',  axs[1][1], 'Closed (Erosion after
    Dilation)')

out_file = 'Square_Morphological.jpg'
show_data(out_file, out_dir='.')
```

### Listing A.4: bspline_demo.py

```python
""" Demonstration of B-splines. """

# author:    stack-overflow user Fnord, comments by Thomas Haslwanter
# date:      April-2021

import numpy as np
import matplotlib.pyplot as plt
import scipy.interpolate as si

from utilities.my_style import set_fonts, show_data


def scipy_bspline(cv, n: int=100, degree: int=3, periodic: bool=False) -> np.
    ndarray:
    """ Calculate n samples on a bspline

    Parameters
    ----------
    cv :   Array of control vertices
    n  :   Number of samples to return
    degree :  Curve degree
    periodic : True - Curve is closed

    Returns
    -------
    spline_data : x/y-values of the spline-curve
    """

    cv = np.asarray(cv)
    count = cv.shape[0]

    # Closed curve
    if periodic:
        kv = np.arange(-degree,count+degree+1)
        factor, fraction = divmod(count+degree+1, count)
        cv = np.roll(np.concatenate((cv,) * factor + (cv[:fraction],)),-1,axis
            =0)
        degree = np.clip(degree,1,degree)

    # Opened curve
    else:
        degree = np.clip(degree,1,count-1)
        kv = np.clip(np.arange(count+degree+1)-degree,0,count-degree)

    # Return samples
    max_param = count - (degree * (1-periodic))
    spl = si.BSpline(kv, cv, degree)
    spline_data = spl(np.linspace(0,max_param,n))

    return spline_data


if __name__ == '__main__':

    cv = np.array([[ 50.,   25.],
        [ 59.,   12.],
        [ 50.,   10.],
        [ 57.,    2.],
        [ 40.,    4.],
        [ 40.,   14.]])
```

```
set_fonts(12)
plt.plot(cv[:,0],cv[:,1], 'o-', label='Control Points')

ax = plt.gca()
ax.set_prop_cycle(None)

# for degree in range(1,7):
for degree in [1, 2, 3]:
    p = scipy_bspline(cv, n=100, degree=degree, periodic=False)
    x,y = p.T
    plt.plot(x, y,  label='Degree %s'%degree)

# Format the plot
plt.legend()
plt.xlim(35, 70)
plt.ylim(0, 30)
plt.xlabel('X')
plt.ylabel('Y')
ax.set_aspect('equal', adjustable='box')

# Show and save the output
out_file = 'bsplines_example.jpg'
show_data(out_file)
```

**Listing A.5: corr_vis.py**

```python
""" Visualization of cross correlation.

The app is working fine, but for some unclear reason, the app crashes the
testing (after it is closed), with the error message
_tkinter.TclError: can't invoke "update" command: application has been destroyed
"""

# author:   Thomas Haslwanter
# date:     April-2021

# Import the required packages
import numpy as np
import matplotlib.pyplot as plt
import seaborn as sns


def corr_vis(x:np.ndarray, y:np.ndarray) -> None:
    """Visualize correlation, by calculating the cross-correlation of two
    signals, one point at a time. The aligned signals and the resulting corss
    correlation value are shown, and advanced when the user hits a key or
    clicks with the mouse.

    Parameters
    ----------
        X : Comparison signal
        Y : Reference signal

    Example
    -------
    x = np.r_[0:2*pi:10j]
    y = sin(x)
    corr_vis(y,x)

    Notes
    -----
    Based on an idea from dpwe@ee.columbia.edu
    """

    Nx = x.size
    Ny = y.size
    Nr = Nx + Ny -1

    xmin = -(Nx - 1)
    xmax = Ny + Nx -1

    # Generate figure and axes
    if not 'fig' in locals():
        fig, axs = plt.subplots(3,1)

    # First plot: Signal 1
    axs[0].plot(range(Ny), y, '-',  label='signal')
    ax = axs[0].axis()
    axs[0].axis([xmin, xmax, ax[2], ax[3]])
    axs[0].xaxis.grid(True)
    axs[0].set_xticklabels(())
    axs[0].set_ylabel('Y[n]')
    axs[0].legend()

    # Pre-calculate limits of correlation output
    axr = [xmin, xmax, np.correlate(x,y,'full').min(),
           np.correlate(x,y,'full').max()]
```

```python
    # Make a version of y padded to the full extent of X's we'll shift
    padY = np.r_[np.zeros(Nx-1), y, np.zeros(Nx-1)]
    Npad = padY.size
    R = []

    # Generate the cross-correlation, step-by-step
    for p in range(Nr):

        # Figure aligned X
        axs[1].cla()
        axs[1].plot(np.arange(Nx)-Nx+p+1, x, '--', label='feature')

        ax = axs[1].axis()
        axs[1].axis([xmin, xmax, ax[2], ax[3]])
        axs[1].xaxis.grid(True)
        axs[1].set_ylabel('X[n-m]')
        axs[1].set_xticklabels(())
        axs[1].legend()

        # Calculate correlation
        # Pad an X to the appropriate place
        padX = np.r_[np.zeros(p), x, np.zeros(Npad-Nx-p)]
        R = np.r_[R, np.sum(padX * padY)]

        # Third plot: cross-correlation values
        axs[2].cla()
        axs[2].plot(np.arange(len(R))-(Nx-1), R,
                    'o-', linewidth=2, color='C1',
                    label='cross-correlation')
        axs[2].axis(axr)
        axs[2].grid(True)
        axs[2].set_xlabel('Steps')
        axs[2].set_ylabel('$R_{xy}[m]$')
        axs[2].legend()

        # Update the plot
        plt.draw()

        # If one exits early, the command "waitforbuttonpress" crashes, and must
        # be caught. Otherwise it produces the error-message
        # _tkinter.TclError: can't invoke "update" command: application has
        # been destroyed try:
            plt.waitforbuttonpress()
        except:
            print('you returned before it was over :(')
            return

    plt.show()


if __name__ == '__main__':
    sns.set_style('ticks')

    # Generate the data used in the book
    signal = np.zeros(20)
    signal[7:10] = 1
    signal[14:17] = 1

    feature = np.zeros(7)
    feature[2:5] = 1

    corr_vis(feature, signal)
```

**Listing A.6: read_zip.py**

```python
"""Get data from MS-Excel files, which are stored zipped on the WWW. """

# author:   Thomas Haslwanter
# date:     April-2021

# Import standard packages
import pandas as pd

# additional packages
import io
import zipfile
from urllib.request import urlopen


def getDataDobson(url: str, inFile: str) -> pd.DataFrame:
    """ Extract data from a zipped-archive on the web. """

    # get the zip-archive
    GLM_archive = urlopen(url).read()

    # make the archive available as a byte-stream
    zipdata = io.BytesIO()
    zipdata.write(GLM_archive)

    # extract the requested file from the archive, as a pandas XLS-file
    myzipfile = zipfile.ZipFile(zipdata)
    xlsfile = myzipfile.open(inFile)

    # read the xls-file into Python, using Pandas, and return the extracted data
    xls = pd.ExcelFile(xlsfile)
    df  = xls.parse('Sheet1', skiprows=2)

    return df


if __name__ == '__main__':
    # Select archive (on the web) and the file in the archive
    url = 'https://work.thaslwanter.at/sapy/GLM.dobson.data.zip'
    inFile = r'Table 2.8 Waist loss.xls'

    df = getDataDobson(url, inFile)
    print(df)

    #input('All done!')
```

**Listing A.7:** fading_astronout.py

```python
"""Add transparency layer to image."""

# author:    Thomas Haslwanter
# date:      June-2020

# Import the required libraries
import numpy as np
import matplotlib.pyplot as plt
from skimage import data

from utilities.my_style import set_fonts, show_data

# Get a color-image
img = data.astronaut()
nrows, ncols = img.shape[:2]

# Make vectors from 1 to 0, with lengths matching the image
alpha_row = np.linspace(1, 0,ncols)
alpha_col = np.linspace(1, 0, nrows)

# Make coordinate-grids
X, Y = np.meshgrid(alpha_row, alpha_col)

# Scale the vector from 0 to 255, and
# let the image fade from top-right to bottom-left
X_Y = np.uint8(X*Y * 255)
X_Y = np.atleast_3d(X_Y)   #make sure the dimensions matches the image

# Add the alpha-layer
img_alpha = np.concatenate( (img, X_Y), axis=2)

plt.imshow(img_alpha)

out_file = 'fading_astronout.png'
show_data(out_file)
```

```python
""" Tips and tricks for interactive work with Matplotlib figures.
Interactive graphs with Matplotlib have haunted me. So here I have collected a
number of tricks that should make interactive use of plots simpler. The
functions below show how to

    - Position figures on the screen (e.g. top left half of display)
    - Pause to display the plot, and proceed automatically after a few sec
    - Proceed on a click, or a keyboard hit
    - Evaluate keyboard inputs
    - Show information on selected data points

based on http://scipy-central.org/item/84/1/simple-interactive-matplotlib-plots
license: Creative Commons Zero (almost public domain) http://scpyce.org/cc0
"""

# author:   Thomas Haslwanter
# date:     April-2021

# Import standard packages
import numpy as np
import matplotlib.pyplot as plt

# additional packages
import tkinter as tk
import plotly.express as px
import pandas as pd
import io

t = np.arange(0,10,0.1)
c = np.cos(t)
s = np.sin(t)


def normalPlot() -> None:
    """Just show a plot. The progam stops, and only continues when the plot is
    closed, either by hitting the "Window Close" button, or by typing "ALT+F4".
    """

    plt.plot(t,s)
    plt.title('Normal plot: you have to close it to continue\n'
            +'by clicking the "Window Close" button, or by hitting "ALT+F4"')
    plt.show()


def positionOnScreen() -> None:
    """Position two plots on your screen.
    This uses the Tickle-backend, which I think is the default on all platforms.
    """

    # Get the screen size
    root = tk.Tk()
    (screen_w, screen_h) = (root.winfo_screenwidth(), root.winfo_screenheight())
    # This gives me a tkinter warning on the next commant. But I found no
    # other way to hide the root window AND continue at the end of the program.
    # Possibly a tkinter bug.
    root.destroy()

    def positionFigure(figure, geometry):
        """Position one figure on a given location on the screen.
        This works for Tk and for Qt5 backends, but may fail on others."""
```

```python
        mgr = figure.canvas.manager
        (pos_x, pos_y, width, height) = geometry
        try:
            # positioning commands for Tk
            position = '{0}x{1}+{2}+{3}'.format(width, height, pos_x, pos_y)
            mgr.window.geometry(position)
        except TypeError:
            # positioning commands for Qt5
            mgr.window.setGeometry(pos_x, pos_y, width, height)


    # The program continues after the first plot
    fig, ax = plt.subplots(1,1)
    ax.plot(t,c)
    ax.set_title('Top Left: Close this one last')

    # Position the first graph in the top-left half of the screen
    topLeft = (0, 0, screen_w//2, screen_h//2)
    positionFigure(fig, topLeft)

    # Put another graph in the top right half
    fig2 = plt.figure()
    ax2 = fig2.add_subplot(111)
    ax2.plot(t,s)
    # I don't completely understand why this one has to be closed first.
    # But otherwise the program gets unstable.
    ax2.set_title('Top Right: Close this one first (e.g. with ALT+F4)')

    topRight = (screen_w//2, 0, screen_w//2, screen_h//2)
    positionFigure(fig2, topRight)

    plt.show()


def showAndPause() -> None:
    """Show a plot only for 2 seconds, and then proceed automatically."""

    plt.plot(t,s)
    plt.title('Don''t touch! I will proceed automatically.')

    plt.show(block=False)
    duration = 2     # [sec]
    plt.pause(duration)
    plt.close()


def waitForInput() -> None:
    """ This time, proceed with a click or by hitting any key """

    plt.plot(t,c)
    plt.title('Click in that window, or hit any key to continue')

    plt.waitforbuttonpress()
    plt.close()


def keySelection() -> None:
    """Wait for user intput, and proceed depending on the key entered.
    This is a bit complex. But None of the versions I tried without
    key binding were completely stable.
    """

    fig, ax = plt.subplots()
```

```python
        fig.canvas.mpl_connect('key_press_event', on_key_event)

        # Disable default Matplotlib shortcut keys:
        keymaps = [param for param in plt.rcParams if param.find('keymap') >= 0]
        for key in keymaps:
            plt.rcParams[key] = ''

        ax.plot(t,c)
        ax.set_title('First, enter a vowel:')
        plt.show()


def on_key_event(event) -> None:
    """Example for keyboard interaction"""

    #print('you pressed %s'%event.key)
    key = event.key

    # In Python 2.x, the key gets indicated as "alt+[key]"
    # Bypass this bug:
    if key.find('alt') == 0:
        key = key.split('+')[1]

    curAxis = plt.gca()
    if key in 'aeiou':
        curAxis.set_title('Well done!')
        plt.pause(1)
        plt.close()
    else:
        curAxis.set_title(key + ' is not a vowel: try again to find a vowel ....
            ')
        plt.draw()


def plotly_demo() -> None:
    """Example for interactive display of information on data.
    To show the name, simply hover the cursor above the data point """

    # Get the data, from a string
    in_data = """name      size      weight
    Peter     175       80
    Paul      180       68
    Mary      165       65"""
    df = pd.read_csv(io.StringIO(in_data), delim_whitespace=True)

    # Plot them, indicating also the 'name' in the hover-display
    fig = px.scatter(df, x="weight", y="size", hover_data=['name'])
    fig.show()

if __name__ == '__main__':
    normalPlot()
    positionOnScreen()
    showAndPause()
    waitForInput()
    keySelection()
    plotly_demo()
```

**Listing A.9: F9_1_fft_sines.py**

```python
""" Example of Fourier Transformation and Power-spectra.

If you want the output to use the LaTeX-formatting, please
    i) make sure that LaTeX is installed properly on your system, and then
    ii) manually set the flag 'latex_installed' in line 22 to 'True'

Also note that the generation of LaTex-formatted figures is rather slow,
since LaTeX has to be launched in the background.
"""

# author:   Thomas Haslwanter
# date:     April-2021

# Import the standard packages
import numpy as np
import matplotlib.pyplot as plt
from scipy import signal
import pandas as pd
import os
from typing import Tuple

# For simplified presentation
from utilities.my_style import set_fonts, show_data

latex_installed = True

if latex_installed:
    import matplotlib
    matplotlib.rcParams['text.usetex'] = True


def generate_data() -> Tuple[np.ndarray, float, np.ndarray, np.ndarray]:
    """ Generate sample data for the FFT-demo
    Signal is a  superposition of three sine waves.

    Returns
    -------
    t : time vector [s]
    dt : sample interval [s]
    sig_with_noise : signal vector, with random noise added
    sig_without_noise : signal vector
    """

    # First set the parameters
    rate = 200      # [Hz]
    duration = 60   # [sec]
    freqs = [3, 7, 20]
    amps = [1, 2, 3]

    # Then calculate the data
    dt = 1/rate
    t = np.arange(0, duration, dt)

    # The clear way of doing it
    sig = np.zeros_like(t)
    for (amp, freq) in zip(amps, freqs):
        omega = 2 * np.pi * freq
        sig += amp * np.sin(omega*t)

    # Add some noise, and an offset
    np.random.seed(12345)
```

```
        offset = 1
        noise_amp = 5
        sig_without_noise = sig + offset
        sig_with_noise = sig_without_noise + noise_amp * np.random.randn(len(sig))

        # Note that the same could be achived with a single line of code.
        # However, in my opinion that is much less clear
        #sig = np.ravel(np.atleast_2d(amps) @ np.sin(2*np.pi * np.c_[freqs]*t)) + \
        #    1 + np.random.randn(len(t))*5

        return (t, dt, sig_with_noise, sig_without_noise)


def power_spectrum(t: np.ndarray, dt: float,
                   sig: np.ndarray,
                   sig_ideal: np.ndarray) -> None:
    """ Demonstrate three different ways to calculate the power-spectrum

    Parameters
    ----------
    t : time [sec]
    dt : sample period [sec]
    sig : sample signal to be analyzed
    sig_ideal : signal without noise
    """

    set_fonts(16)

    fig, axs = plt.subplots(1,2, figsize=(10, 5))

    if latex_installed:
        txt = '$\displaystyle signal=offset + \sum_{i=0}^{2} a_i*sin(\omega_i*t)
                + noise$'

        label = '$|FFT|\; ()$'
        label2 = '$|FFT|^2  ()$'
    else:
        txt = 'signal = offset + sum(i=0:2) a_i*sin(omega_i*t)'
        label = '|FFT| ()'
        label2 = '|FFT|^2 ()'

    # From a quick look we learn - nothing
    axs[0].plot(t, sig, lw=0.7, label='noisy')
    axs[0].plot(t, sig_ideal, ls='dashed', lw=2, label='ideal')
    axs[0].set_xlim(0, 0.4)
    axs[0].set_ylim(-15, 25)
    axs[0].set_xlabel('Time (s)')
    axs[0].set_ylabel('Signal ()')
    axs[0].legend()
    axs[0].text(0.2, 26, s=txt, fontsize=16)

    # Calculate the spectrum by hand
    fft = np.fft.fft(sig)
    print(fft[:3])
    fft_abs = np.abs(fft)

    # The easiest way to calculate the frequencies
    freq = np.fft.fftfreq(len(sig), dt)

    axs[1].plot(freq, fft_abs)
    axs[1].set_xlim(0, 35)
    axs[1].set_xlabel('Frequency (Hz)')
```

```
        axs[1].set_ylabel(label)

        axs[1].set_yticklabels([])
        #plt.show()
        show_data('FFT_sines.jpg')

        # Also show the double-sided spectrum
        fig, ax = plt.subplots(1,1)
        ax.plot(fft_abs)
        #ax.set_xlim(0, 35)
        ax.set_xlabel('Points')
        ax.set_ylabel(label)
        plt.tight_layout()
        show_data('FFT_doublesided.jpg')

        # With real input, the power spectrum is symmetrical and we only one half
        fft_abs = fft_abs[:int(len(fft_abs)/2)]
        freq = freq[:int(len(freq)/2)]

        # The power is the norm of the amplitude squared
        Pxx = fft_abs**2
        # Showing the same data on a linear and a log scale
        fig, axs = plt.subplots(2,1, sharex=True)
        axs[0].plot(freq, Pxx)
        axs[0].set_ylabel('Power (linear)')
        axs[1].semilogy(freq, Pxx)
        axs[1].set_xlabel('Frequency (Hz)')
        axs[1].set_ylabel('Power (dB)')
        show_data('FFT_sines_power_lin_log.jpg')

        # Periodogram and Welch-Periodogram
        f_pgram, P_pgram = signal.periodogram(sig, fs = 1/dt)
        f_welch, P_welch = signal.welch(sig, fs = 100, nperseg=2**8)

        fig, axs = plt.subplots(2, 1, sharex=True)

        axs[0].semilogy(f_pgram, P_pgram, label='periodogram')
        axs[1].semilogy(f_welch, P_welch, label='welch')

        axs[0].set_ylabel('Spectral Density (dB)')
        axs[0].legend()
        axs[0].set_ylim(1e-4, 1e3)
        axs[1].set_xlabel('Frequency (Hz)')
        axs[1].set_ylabel('Spectral Density (dB)')
        axs[1].legend()
        show_data('FFT_sines_periodogram.jpg')


    if __name__ == '__main__':
        data = generate_data()

        power_spectrum(*data)
        # Equivalent to:
        #power_spectrum(data[0], data[1], data[2], data[3])
```

# Appendix B

# Solutions

## B.1  Solutions to Introduction

The answers to Exercise 1.1 are:

a) $\begin{pmatrix} 1 & 2 \end{pmatrix} \cdot \begin{pmatrix} 3 \\ 4 \end{pmatrix} = 11$

b) $\begin{pmatrix} 3 \\ 4 \end{pmatrix} \cdot \begin{pmatrix} 1 & 2 \end{pmatrix} = \begin{bmatrix} 3 & 6 \\ 4 & 8 \end{bmatrix}$

c) $\begin{pmatrix} 1 & 2 \end{pmatrix} \cdot \begin{pmatrix} 3 & 4 \end{pmatrix} \rightarrow$ "Error using $*$ . Incorrect dimensions for matrix multiplication."

## B.2  Solutions to Python

### 1. Translation

**Listing B.1: S2_translation.py**

```python
""" Solution to Exercise 'Translation', Chapter 'Python' """

# author:    Thomas Haslwanter
# date:      April-2021

# Import the required packages
import numpy as np
import matplotlib.pyplot as plt

# Define the original points
p_0 = [0,0]
p_1 = [2,1]

# Combine them to an array
array = np.array([p_0, p_1])
print(array)

# Translate the array
translated = array + [3,1]
print(translated)

# Plot the data
plt.plot(array[:,0], array[:,1], label='original')
plt.plot(translated[:,0], translated[:,1], label='translated')
```

T. Haslwanter, *An Introduction to Hands-on Signal Analysis with Python*,

```python
# Format and show the plot
plt.xlabel('X')
plt.ylabel('Y')
plt.legend()
plt.show()
```

## 2. **Rotation**

```python
""" Solution to Exercise 'Rotation', Chapter 'Python' """

# author:   Thomas Haslwanter
# date:     April-2021

# Import the required packages
import numpy as np
import matplotlib.pyplot as plt


def rotate_me(in_vector:np.ndarray, alpha:float) -> np.ndarray:
    """Function that rotates a vector in 2 dimensions

    Parameters
    ----------
    in_vector : vector (2,) or array (:,2)
                vector(s) to be rotated
    alpha : rotation angle [deg]

    Returns
    -------
    rotated_vector : vector (2,) or array (:,2)
                rotated vector

    Examples
    --------
    perpendicular = rotate_me([1,2], 90)

    """

    alpha_rad = np.deg2rad(alpha)
    R = np.array([[np.cos(alpha_rad), -np.sin(alpha_rad)],
    [np.sin(alpha_rad), np.cos(alpha_rad)]])
    return R @ in_vector


if __name__ == '__main__':

    vector = [2,1]
    # Draw a green line from [0,0] to [2,1]
    plt.plot([0,vector[0]], [0, vector[1]], 'g', label='original')

    # Coordinate system
    plt.hlines(0, -2, 2, linestyles='dashed')
    plt.vlines(0, -2, 2, linestyles='dashed')

    # Make sure that the x/y dimensions are equally drawn
    cur_axis = plt.gca()
    cur_axis.set_aspect('equal')
```

```python
    # Rotate the vector
    rotated = rotate_me(vector, 25)
    plt.plot([0, rotated[0]], [0 ,rotated[1]],
             label='rotated',
             color='r',
             linewidth=3)

    plt.legend()
    plt.show()
```

## 3. Taylor

**Listing B.3: S2_Taylor.py**

```python
""" Solution to Exercise 'Taylor', Chapter 'Python' """

# author:   Thomas Haslwanter
# date:     April-2021

# Import the required packages
import numpy as np
import matplotlib.pyplot as plt
from typing import Tuple


def approximate(angle:np.ndarray) -> Tuple[np.ndarray, np.ndarray]:
    """Function that calculates a second order approximation to sine and cosine

    Parameters
    ----------
    angle : angle [deg]

    Returns
    -------
    approx_sine :  approximated sine
    approx_cosine :  approximated cosine

    Examples
    --------
    alpha = 0.1
    sin_ax, cos_ax = approximate(alpha)

    Notes
    -----
    Input can also be a single float

    """

    sin_approx = angle
    cos_approx = 1 - angle**2/2

    return (sin_approx, cos_approx)


if __name__ == '__main__':
    limit = 50          # [deg]
    step_size = 0.1     # [deg]
```

```python
    # Calculate the data
    theta_deg = np.arange(-limit, limit, step_size)
    theta = np.deg2rad(theta_deg)
    sin_approx, cos_approx = approximate(theta)

    # Plot the data
    plt.plot(theta_deg, np.column_stack((np.sin(theta), np.cos(theta))), label='
        exact')
    plt.plot(theta_deg, np.column_stack((sin_approx, cos_approx)),
             linestyle='dashed',
             label='approximated')
    plt.legend()
    plt.xlabel('Angle [deg]')
    plt.title('sine and cosine')
    out_file = 'approximations.png'
    plt.savefig(out_file, dpi=200)
    print(f'Resulting image saved to {out_file}')

    plt.show()
```

## 4. First Steps with Pandas

**Listing B.4: S2_first_steps.py**

```python
""" Solution to Exercise 'First Steps with Pandas', Chapter 'Python' """

# author:   Thomas Haslwanter
# date:     April-2021

# Import the required packages
import numpy as np
import matplotlib.pyplot as plt
import pandas as pd

# Set the parameters
rate =   10
start, stop = 0, 10
freq = 1.5
out_file = 'out.txt'

# Calculate the data
dt = 1/rate
omega = 2*np.pi*freq

x = np.arange(start, stop, dt)
y = np.sin(omega*x)
z = np.cos(omega*x)

# Put them into a pandas DataFrame
df = pd.DataFrame({'Time': x, 'YVals': y, 'ZVals': z})

# Show the head
print(df.head())

# Extract rows 10-15. Be careful with the last row!
data = df[ ['YVals', 'ZVals'] ][10:16]

# Write them to the out-file
data.to_csv(out_file)
print(f'Data have been written to {out_file}')
```

# B.3 Solutions to Data Input

## 1. Reading in Data

---

**Listing B.5: S3_data_gen.py**

```python
""" Generation of data for Chapter 'Data Input'

It shows how to generate
- formatted text-strings
- CSV files                        -> 'data.csv'
- otherwise formmated TXT-files  -> 'data_tab.txt', 'data_modified.txt'
- Excel files                      -> 'data.xls'
- Matlab files                     -> 'data.mat'
- Binary data                      -> 'data.raw'

One may have to install the package "xlwt" for this solution to run.
"""


# author:   Thomas Haslwanter
# date:     April-2021

# Import the required packages
import numpy as np
import matplotlib.pyplot as plt
import pandas as pd


def save_txt(df: pd.DataFrame, out_file='data.csv') -> None:
    """ Save data to ASCII-format

    Parameters
    ----------
    df : Input data
    out_file : Output file; two other ASCII-files with same stem
               are also generated
    """

    # Saves the data to CSV-format, which means by default
    # - Separated by a comma
    # - With a column name
    # - With a running index on the left side
    df.to_csv(out_file)

    # Always let the user know when you generate a new file!!
    # If you use Python >3.7, you can use the "format-strings"
    print(f'Data have been saved in CSV-format to {out_file}')

    # For earlier versions of Python you have to use
    # print('Data have been saved in CSV-format to {0}'.format(out_file))

    # Simple file, tab-separated, no header, no index
    simple_file = out_file.replace('.csv', '_tab.txt')
    df.to_csv(simple_file, sep='\t', header=False, index=False)
    print(f'Data have been saved to {simple_file}')

    # Show how to add a file-header to an existing text file
    with open(out_file, 'r') as original:
        text = original.read()
```

```python
    modified_file = out_file.replace('.csv', '_modified.txt')
    with open(modified_file, 'w') as modified:
        modified.write('This file was generated on Sept 19\n')
        modified.write(f'Sampling rate: {rate} [Hz]\n')
        modified.write(text)
    print(f'A file header has been added to {out_file}, and the new file saved
        as {modified_file}')


def save_xls(df: pd.DataFrame, out_file='data.xls') -> None:
    """ Save data to MS Excel-format

    Parameters
    ----------
    df : Input data
    out_file : Name of output file
    """

    # Save data to Excel-format
    df.to_excel(out_file, index=False)
    print(f'Data have been saved in MS-Excel format, to {out_file}')


def save_matlab(t: np.ndarray, data: np.ndarray, out_file='data.mat') -> None:
    """ Save data to Matlab-format

    Parameters
    ----------
    t : Time-values [sec]
    data : sine-wave
    out_file : Name of output file
    """
    # To save data to Matlab-format, we need the package "scipy.io", ...
    from scipy.io import savemat

    # ... and we have to put the data into a Python-dictionary
    # For this example, I add an information-text, and format the data as a
        matrix
    data_mat = np.column_stack( (t, data) )
    data_dict = {'info':'These are demo-data, showing a sine-wave',
                 'data': data_mat}

    savemat(out_file, data_dict)
    print(f'Data have been saved in Matlab format, to {out_file}')


def generate_binary(out_file='data.raw') -> None:
    """"Generate binary data, with an ASCII-header with 256 byte, and
    three columns of data.

    Parameters
    ----------
    out_file : name of outfile
    """

    # To generate a more interesting signal, I produce a "chirp"
    from scipy.signal import chirp

    # Set the parameters
    length_header = 256      # byte

    # Generate a dummy header text
    txt = '''This is the header.
```

```
    It has a length of 'length_header' byte. After the text,
    it is padded with whitespaces.'''
    out_txt = txt + ' '*(length_header-len(txt))

    # Write it to the out_file
    fh = open(out_file, 'wb')
    fh.write(out_txt.encode())

    # Generate some data
    t = np.arange(0,20, 0.1)
    x = np.sin(t)
    y = chirp(t, 3, np.max(t), 0.01)
    data = np.column_stack((t, np.sin(t), y*t))

    # Also write them to a file
    fh.write(data.tobytes())
    fh.close()
    print(f'Data have been written in binary form to {out_file}')


if __name__ == '__main__':

    # Set the parameters for a sine wave
    rate = 50
    freq = 2
    duration  = 4
    amp = 5
    noise_amp = 0.8

    # Calculate the sine-values
    delta = np.deg2rad(15)
    dt = 1/rate
    omega = 2*np.pi*freq

    t = np.arange(0, duration, dt)
    data = amp * np.sin(omega*t + delta) + noise_amp * np.random.randn(len(t))

    # Put them in a pandas-DataFrame, for easier text output
    df = pd.DataFrame({'t':t, 'values':data})

    # Show how to generate a formatted string in Python
    print(f'The first time-sample is {t[0]:5.3f}, and the first data-value is {
        data[0]:5.3f}\n')

    # Generate the output files
    save_txt(df)
    save_xls(df)
    save_matlab(t, data)
    generate_binary()
```

## 2. Modifying Text Files: Imaginary Numbers

**Listing B.6: S3_data_read.py**

```
""" Solution to Exercise 'Reading in Data', Chapter 'Data Input' """

# author:   Thomas Haslwanter
# date:     April-2021
```

```
# Import the standard packages
import numpy as np
import matplotlib.pyplot as plt
import pandas as pd

# ------------ data.csv --------------------------------
# The name "df" indicates a Pandas-DataFrame
in_file = 'data.csv'
df = pd.read_csv(in_file, index_col=0)
print(df.head())
print(df.tail())

# ------------ data_tab.txt ----------------------------
in_file = 'data_tab.txt'
df = pd.read_csv(in_file, sep='\t', header=None)
print(df.head())
print(df.tail())

# ------------ data_modified.txt -----------------------
in_file = 'data_modified.txt'
df = pd.read_csv(in_file, sep=',', header=2, index_col=0)
df.plot('t', 'values')
plt.show()

# ------------ data.xls --------------------------------
in_file = 'data.xls'
df = pd.read_excel(in_file)
print(df.head())

# ------------ data.mat --------------------------------
from scipy.io import loadmat
in_file = 'data.mat'
data_dict = loadmat(in_file)
print(data_dict['info'])
data = data_dict['data']
plt.plot(data[:,0], data[:,1])
plt.show()
```

## 3. Mixed Inputs

**Listing B.7: S3_imaginary.py**

```
""" Solution to Exercise 'Modifying Text Files', Chapter 'Data Input' """

# author:   Thomas Haslwanter
# date:     April-2021

# Import the required packages
import numpy as np
import matplotlib.pyplot as plt
import pandas as pd
import os

# Set the required parameters
data_dir = '../../data'
file_name = 'imaginary.txt'
in_file = os.path.join(data_dir, file_name)
out_file = 'imaginary_out.txt'
```

```
# Get the data
df = pd.read_csv(in_file, delim_whitespace=True)

# Add radius and angle as new columns
df['Radius'] = np.sqrt(df.Real**2 + df.Imaginary**2)
df['Angle [rad]'] = np.arctan2(df.Imaginary, df.Real)

# Make sure all columns are floats, and write to the new file
df = df.astype('float')
df.to_csv(out_file, sep='\t', index=None, float_format='%5.3f')
print(f'Modified data saved to {out_file}')
```

## 4. Binary Data

**Listing B.8: S3_read_binary.py**

```
""" Solution to Exercise 'Binary Data', Chapter 'Data Input'

Read in binary data, with an 256 byte ASCII-header.
"""

# author:    Thomas Haslwanter
# date:      April-2021

# Import the required packages
import numpy as np
import matplotlib.pyplot as plt
import array


# Set the parameters
data_file = 'data.raw'
length_header = 256      # byte

# Approach #1: -------------------------------

# Read and show the file header
fh = open(data_file, 'rb')
txt = fh.read(length_header).decode()
print(txt)

# Read all the binary data
bin_data = fh.read()

# Interpret them as 'double', and reshape them to an ndarray
double = 8   # byte
num_cols = 3
num_data = int(len(bin_data)/(num_cols*double))    # needs to be an integer
values = array.array('d', bin_data)
mat = np.reshape(values, (num_data,-1))

# Assign the three columns to the variables (t,x,y)
t,x,y = mat.T

# Plot the data
fig, axs = plt.subplots(2,1)
axs[0].plot(t, x)
axs[1].plot(t, y)
axs[0].set_title('Retrieved Data')
plt.show()
```

```
# Approach #2: -------------------------------

# Define a structured array
dt = np.dtype([('t', 'd'), ('x', 'd'), ('y', 'd')])

# Position the file pointer after the header
fh.seek(256)

# Read the data in as a structured array
structured_array = np.fromfile(fh, dtype=dt)

# Convert that array to a numpy-ndarray
mat = np.array(structured_array.tolist())

# Plot the data again
fig, axs = plt.subplots(2,1)
axs[0].plot(t, x)
axs[1].plot(t, y)
axs[0].set_title('Retrieved Data, Version 2')
plt.show()
```

# B.4    Solutions to Data Display

## 1. Plotting Data

Listing B.9: S4_plot_data.py

```python
""" Solution to Exercise 'Plotting Data', 'Data Display' """

# author:    Thomas Haslwanter
# date:      April-2021

# Import the required packages
import numpy as np
import matplotlib.pyplot as plt


def plot_data(num_cycles:float=1, freq:float=1) -> None:
    """"Calculate and plot a sine wave.

    Parameters
    ----------
    num_cycles: Number of cycles
    freq : Frequency of oscillation [Hz]

    Examples
    --------
    plot_data(5, 0.3)
    """

    # Set the parameters
    amp = 1
    rate = 100        # [Hz]
    noise_amp = 0.5

    # Calculate the data
    dt = 1/rate
    omega = 2*np.pi*freq
    t_cycle = 1/freq

    t = np.arange(0, num_cycles*t_cycle, dt)
    x = amp * np.sin(omega*t) + noise_amp*np.random.randn(len(t))

    # Plot the data
    plt.plot(t, x);
    plt.xlabel('Time [s]');
    plt.ylabel('Signal');
    plt.title('Noisy Sine');

    plt.show()


if __name__ == '__main__':
    plot_data(2, 0.3)
```

## 2. Modifying Figures

Listing B.10: S4_modify_figures.py

```python
""" Solution to Exercise 'Modifying Figures', Chapter 'Data Display' """
```

```python
# author:    Thomas Haslwanter
# date:      April-2021

# Import the required packages
import numpy as np
import matplotlib.pyplot as plt

# Set the positions of the figure elements
yi = 0.8
xi = 2*np.pi + np.arcsin(yi)
dx = 3
dy = 0.5

out_file = 'drawing.jpg'

# Generate the sine-wave
t = np.arange(0, 10, 0.1)
x = np.sin(t)

# Plot it
plt.plot(t,x)
plt.axhline(yi, ls='dotted')

# Annotate it
plt.annotate('This is\nnot funny!',
             xy = (xi,yi),
             xytext = (xi-dx, yi-dy),
             arrowprops=dict(facecolor='black', shrink=0.05) )

# Save JPG-file
pil_kwargs = {'quality': 90}
plt.savefig(out_file, dpi=200, pil_kwargs=pil_kwargs)

# Save the same file in SVG-format
svg_file = out_file.replace('jpg', 'svg')
plt.savefig(svg_file)
print(f'Saved {out_file} and {svg_file}')
plt.show()
plt.close()

# Now re-generate the plot in a "funny"-style, and save the file
# with "_funny" added to the JPG filename
with plt.xkcd():
    # Plot it
    plt.plot(t,x)
    plt.axhline(yi, ls='dotted')

    # Annotate it
    plt.annotate('This is\nfunny!',
                 xy = (xi,yi),
                 xytext = (xi-dx, yi-dy),
                 arrowprops=dict(facecolor='black', shrink=0.05) )

    funny_file = out_file.replace('.jpg', '_funny.jpg')
    plt.savefig(funny_file, dpi=200, pil_kwargs=pil_kwargs)
    print(f'... and also saved {funny_file} ;)')
    plt.show()
```

# B.5 Solutions to Data Filtering

1. **Integration as IIR-filter**
A necessary condition for the implementation of integration as an IIR-filter is that all $\Delta x_i$ are equal.

**cumsum**
With this in mind, implementation of `np.cumsum` as an IIR filter is straightforward. The first few values of `cumsum` are

$$y_0 = x_0$$
$$y_1 = x_0 + x_1 = y_0 + x_1$$
$$y_2 = x_0 + x_1 + x_2 = y_1 + x_2$$
$$\vdots$$
$$y_i = y_{i-1} + x_i$$

Bringing all the $y's$ onto the left side, and all the $x's$ to the right side (Eq. 5.12) gives us the parameters for the command `scipy.signal.lfilter(b,a,x)`:

$$\mathbf{a} = [1, -1]$$
$$\mathbf{b} = [1]$$

**cumtrapz**
Writing down the values of `scipy.integrate.cumtrapz` in the same way gives

$$y_0 = 0$$
$$y_1 = \frac{x_0 + x_1}{2}$$
$$y_2 = y_1 + \frac{x_1 + x_2}{2}$$
$$\dots$$
$$y_i = y_{i-1} + \frac{x_1}{2} + \frac{x_{i-1}}{2}$$

giving us the `lfilter` parameters

$$\mathbf{a} = [1, -1]$$
$$\mathbf{b} = [0.5, 0.5]$$

Listing B.11: S5_integral_as_iir.py

```
""" Solution Ex. 'Integration as  IIR-filte, Chapter 'Data Filtering' """

# author:    Thomas Haslwanter
# date:      April-2021
```

```python
# Import the basic packages
import numpy as np
import matplotlib.pyplot as plt
import pandas as pd

from scipy.signal import lfilter
from scipy import integrate
from collections import namedtuple

a = [1, -1]
b = {'cum_trapz': np.r_[0.5, 0.5],
     'cum_sum':   [1]}

Results = namedtuple('Results', ['calculated', 'filtered'])

# First show effect with 0:5
x = np.arange(6, dtype=float)

print('cum_sum: -----------')
cum_sum = Results(np.cumsum(x),
                  lfilter(b['cum_sum'], a, x))
print(cum_sum)

print('cum_trapz: -----------');
cum_trapz = Results(integrate.cumtrapz(x),
                    lfilter(b['cum_trapz'], a, x) )
print(cum_trapz)

# For the quadratic fit, use a sine-wave
dt = 0.1     # [sec]
t = np.arange(0, 3*np.pi, dt)

# Calculate and show exact values of sine and its integral
si = np.sin(t)
co = np.cos(t)

plt.plot(t, si, label='sin')
plt.plot(t, 1-co, label='1-cos')

# Approximal integrals
integral = {}
integral['cum_sum'] = np.cumsum(si) * dt
integral['cum_trapz'] = integrate.cumtrapz(si) * dt

plt.plot(t, integral['cum_sum'], '-*', label='cum_sum')
plt.plot(t[1:], integral['cum_trapz'], '-o', label='cum_trapz')

plt.legend()
plt.show()

# Show the first ten numbers
print('cosine (exact) -----------')
print( 1-co[:10] )
print('sine integral: cum_sum ------------')
print( integral['cum_sum'][0:10:2] )
print('sine integral: cum_trapz ------------')
print( integral['cum_trapz'][0:10:2] )
```

## 2. Differentiation of Noisy Data

Listing B.12: S5_noisy_sine.py

```python
""" Solution Exercise 'Smoothing and Differentiation of Noisy Data',
    Chapter 'Data Filtering' """

# author:   Thomas Haslwanter
# date:     April-2021

# Import the required packages
import numpy as np
import matplotlib.pyplot as plt
from scipy.signal import savgol_filter

# Set the parameters
amp = 1
freq = 0.3
num_cycles = 5
rate = 100
noise_amp = 0.5

# Calculate the noisy sine-wave
dt = 1/rate
omega = 2*np.pi*freq
tCycle = 1/freq
t = np.arange(0, num_cycles*tCycle, dt)
x = amp * np.sin(omega*t) + noise_amp*np.random.randn(len(t))

# Smooth the noisy data, and check the result visually
filtered  = savgol_filter(x, 31, polyorder=2)
plt.plot(t, x, label='original')
plt.plot(t, filtered, label='smoothed')
plt.xlabel('Time (s)')
plt.ylabel('Signal')
plt.legend()
plt.show()

# Calculate and plot the first derivative of the noisy sine-wave
win_sizes = [31, 51, 101]
gray_levels = [0.85, 0.65, 0.2]

for (grayLevel, winSize) in zip(gray_levels, win_sizes):
    filtered = savgol_filter(x, winSize, 2, 1, dt)
    gray = grayLevel * np.ones(3)
    plt.plot(t, filtered, color=gray)

plt.title('Differentiated Signal')
plt.legend(( f'winSize: {win_sizes[0]}',
        f'winSize: {win_sizes[1]}',
        f'winSize: {win_sizes[2]}'))
plt.show()
```

## 3. Band-pass Filter

Listing B.13: S5_BandPass.py

```python
""" Solution Exercise 'Bandpass', Chapter 'Data Filtering' """

# author:   Thomas Haslwanter
```

```python
# date:        April-2021

# Import the required packages
import numpy as np
import matplotlib.pyplot as plt
from scipy import signal

# Set the parameters
freqs = np.c_[[2, 30, 400]]
amps = np.r_[0.5, 1, 0.1]
rate = 5000
duration = 2

# Note that we want the "freq" as an numpy array, so we can normalize it
#    afterwards
bandpass = {'freq': np.r_[10, 100],
            'order': 3}

# Calculate the signal
dt = 1/rate
t = np.arange(0, duration, dt)
t = np.atleast_2d(t)
nyq = rate/2    # Nyquist frequency
x = amps @ np.sin(2*np.pi * freqs @ t)    #If you don't believe it, do it
#    explicitly

# Band-pass filter the data
[b,a] = signal.butter(bandpass['order'], bandpass['freq']/nyq, 'bandpass')
filtered = signal.lfilter(b,a,x)

# Show the data
fig, axs = plt.subplots(2,1, sharex=True)
# The "sharex" allows you to zoom in simultaneously on both axes

# The "flatten" turns a 2d-column array with one column into a plain vector
axs[0].plot(t.flatten(), x)
axs[0].set_ylabel('Rawdata')
axs[0].XTickLabels = []      # Only show the necessary information, no redundancy

axs[1].plot(t.flatten(), filtered)
axs[1].set_xlabel('Time [s]');
axs[1].set_ylabel('Filtered');

plt.show()
```

## 4. Exponential Averaging Filter

**Listing B.14: S5_exp_avg.py**

```python
""" Solution Ex. 'Exponential Averaging Filter', Chapter 'Data Filtering' """

# author:   Thomas Haslwanter
# date:      April-2021


# Import the required packages
import numpy as np
import matplotlib.pyplot as plt
from scipy import signal
```

```python
def applyLeakyIntegrator(alpha:float, x:np.ndarray) -> np.ndarray:
    """
    Parameters
    ----------
        alpha : Decay rate of leaky integrator
        x : ndarry (N,)
            input data

    Return
    ------
    filtered : numpy vector (N,)
            Filtered data

    Example
    -------
    filtered = applyLeakyIntegrator(0.3, np.random.randn(200))
    """

    b = [alpha]
    a = [1, -(1-alpha)]
    filtered = signal.lfilter(b, a, x)

    return filtered


if __name__ == '__main__':
    # Prepare the input step, and the time-axis for plotting
    x = np.zeros(50)
    x[10:] = 1
    t = np.arange(len(x))

    # Plot the response, for different values of alpha
    alphas = [0.1, 0.2, 0.5, 0.9]
    colors = 'brgk'

    for (alpha, color) in zip(alphas, colors):
        plt.plot(t, applyLeakyIntegrator(alpha, x), color=color)

    plt.xlabel('Time')
    plt.ylabel('Step-responses')
    plt.legend(alphas)

    plt.show()
```

# B.6    Solutions to Event- and Feature-Finding

## 1. Event Finding

**Listing B.15: S6_events.py**

```python
""" Solution Exercise 'Event Finding', Chapter 'Events' """

# author:   Thomas Haslwanter
# date:     April-2021

# Import the required packages
import numpy as np
import matplotlib.pyplot as plt
from scipy import signal
import os

# Generate the data
in_dir = '../../data'
in_file = 'S6_1_data.npz'
full_in_file = os.path.join(in_dir, in_file)
data = np.load(full_in_file)

data_dict = dict(data)
sig = data_dict.pop('signal', None)  # Get the 'signal' key
features = data_dict                 # Assign the remaining dictionary to '
    features'

# Find where the features occur in the signal
found_locations = dict()      # Empty dictionary
pattern_names = features.keys()

for pattern in pattern_names:
    # Calculate the cross correlation between signal and pattern
    r = signal.correlate(sig, features[pattern])
    lag = np.arange(len(r)) - len(features[pattern])

    # Find the maxima of that cross correlation, by ...
    # ... first finding the approximate matches ...
    threshold = 0.75 * np.max(r)
    approxLocations = (r > threshold)*1
    local_starts = np.where(np.diff(approxLocations) ==  1)[0]
    local_ends   = np.where(np.diff(approxLocations) == -1)[0]

    # ... and then the local maxima
    maxLocs = []
    for (start, end) in zip(local_starts, local_ends):
        locMax = r[start:end].argmax()
        # the "+1" is due to the fact that the "diff" (above) is one shorter
            than the input
        maxLocs.append( lag[start+locMax] + 1)

    found_locations[pattern] = maxLocs

print('Found Locations:')
print(found_locations)
```

## 2. Synchronization

---

```python
""" Solution Exercise 'Synchronization, Chapter 'Events'

Synchronize the acceleration measurements from two datasets
The information about the measurement units is taken from the column names
"""

# author:   Thomas Haslwanter
# date:     April-2021

# Import the required packages
import numpy as np
import matplotlib.pyplot as plt
import pandas as pd
import os

# Get the data
data_dir = '../../data'
in_mobile = 'mobile_phone.txt'
in_imu = 'ngimu.txt'
rate_sync = 100          # [Hz] for synchronized data

# ... mobile phone data
os.chdir(data_dir)
mobile = pd.read_csv(in_mobile,
                     sep='\t',
                     names=['time', 'acc_x', 'acc_y', 'acc_z', 'total'],
                     dtype='float',
                     skiprows=1,
                     decimal=',')

# ... imu data
# Make sure that the 'time' column and the 'acceleration' columns have the same
# name as the ones from the mobile phone, so they can be processed in a loop
imu = pd.read_csv(in_imu)
imu.columns = ['time',
               'w_x', 'w_y', 'w_z',
               'acc_x', 'acc_y', 'acc_z',
               'b_x', 'b_y', 'b_z',
               'p']
# For the imus-values, the acceleration is given in units of 'g'
imu.iloc[:, 4:7] *= 9.81

# Trigger on the 'total' acceleration. For the 'mobile', we already have it
imu_acc = imu.filter(regex='acc*')
imu['total'] = np.sqrt(np.sum(imu_acc**2, axis=1))

ip_ed = []        # store the interpolated data ...
for sensor in ['mobile', 'imu']:
    df = eval(sensor)
    df_new = pd.DataFrame()         # ... in a DataFrame
    start = np.argmax(df.total)     # Point of max acc
    t_int = np.arange(df.time[start], df.time.iloc[-1], 1/rate_sync)

    df_new['time']  = t_int-t_int[0]    # Set this point to '0'
    # For (x/y/z)
    df_acc = df.filter(regex='acc*')
    for col in df_acc.columns:
        df_new['_'.join([sensor, col])] = np.interp(t_int, df.time, df_acc[col])
    ip_ed.append(df_new)
```

```python
# Chop off the longer one
t_max = np.max([ip_ed[0].time.iloc[-1], ip_ed[1].time.iloc[-1]])
for sensor in ip_ed:
    sensor = sensor.drop(sensor[sensor.time > t_max].index)

ip_ed[1] = ip_ed[1].drop(columns='time') # We only need one 'time' column

# Combine the interpolated values in one DataFrame
synced = pd.concat(ip_ed, axis=1)
out_file = 'synchronized.txt'
synced.to_csv(out_file)
print(f'Synchronized data saved to {out_file}' \
      + f', with a sample rate of {rate_sync} Hz.')
```

## 3. Analyze EMG-data

**Listing B.17: S6_emg.py**

```python
""" Solution Exercise 'EMG', Chapter 'Events' """

# author:   Thomas Haslwanter
# date:     April-2021

# Import the required packages
import numpy as np
import matplotlib.pyplot as plt
import pandas as pd
import os
from scipy import signal

# Set the parameters
in_dir = r'..\..\data'
file_name = 'Shimmer3_EMG_calibrated.csv'

# Get the data
in_file = os.path.join(in_dir, file_name)
df = pd.read_csv(in_file, skiprows=3, header=None, delim_whitespace=True)
df.columns = ['date', 'abs_time', 'cal', 'emg_1', 'emg_2']

# Either use a sample-rate of 512 Hz (from experimental protocol)
rate = 512

# or get the time from the time-stamps
df.abs_time = pd.to_datetime(df.abs_time)
df['time'] = df.abs_time - df.abs_time[0]
# convert from pandas format (nano-sec) into sec
df['t_sec'] = df.time.to_numpy(dtype=float)/1e9

# Show the original data
df.plot('t_sec', 'emg_1')
plt.ylabel('Signal (mV)')
plt.show()

# High-pass filter the data, to eliminate drifts
b, a = signal.butter(5, 1/(rate/2), 'high')
df['filtered'] = signal.lfilter(b, a, df.emg_1)

# Smooth the absolute value, to obtain the level of muscle activation
df['smoothed'] = signal.savgol_filter(np.abs(df.filtered), polyorder=3,
        window_length=101)
```

```python
# Find onset and offset of muscle activations
threshold = 0.05
activity = df.smoothed > threshold

onset = np.where(np.diff(activity*1)==1)[0]
offset = np.where(np.diff(activity*1)==-1)[0]

# During the first ca. 4 sec we have startup-artifacts in the filtered data
# so we eliminate those events
onset = onset[onset>2000]
offset = offset[offset>2000]

# Make sure that we start with an onset ...
if onset[0] > offset[0]:
    offset = offset[1:]
# ... and end with an offset
if offset[-1] < onset[-1]:
    onset = onset[:-1]

assert(len(onset)==len(offset))

# Eliminate short contractions, as they may be artifacts
min_interval = 0.5
onsets = []
offsets = []
for (start, stop) in zip(onset, offset):
    dt = (stop-start)/rate
    if dt > min_interval:
        onsets.append(start)
        offsets.append(stop)

# Convert onsets and offsets from lists to arrays
onsets = np.array(onsets)
offsets = np.array(offsets)

# Show the data
df.plot('t_sec', 'smoothed')
plt.plot(np.r_[onsets]/rate, np.zeros_like(onsets), 'ro')
plt.plot(np.r_[offsets]/rate, np.zeros_like(offsets), 'g*')
plt.show()

# And display the mean contraction time
print('The mean contraction time is' +
    f'{np.mean(offsets - onsets)/rate:5.2f} (sec)')
```

## 4. Heart Rate Variability

**Listing B.18: S6_hrv.py**

```python
""" Solution Exercise 'Heart Rate Variability', Chapter 'Events'

Estimation of hear beat variability
To read in the data, you need the package 'wfdb'
"""

# author:   Thomas Haslwanter
# date:     June-2020

import numpy as np
import matplotlib.pyplot as plt
```

```
import os
import wfdb

# Set the parameters
data_dir = '../../data'
in_file = 'rec_1'
rate = 500          # [Hz]
nn = 10             # calculate HRV over nn heart beats

# Import the data
os.chdir(data_dir)
sig, fields = wfdb.rdsamp(in_file)
ecg = sig[:,0]
plt.plot(ecg)

# Set the threshold for R-detection
#threshold = plt.ginput(1)
threshold = 0.4

# Find the locations where the signal exceeds the threshold
high = ecg>threshold
onset = np.where(np.diff(high*1.)==1)[0]

# Times between two heartbeats
dt = 1/rate
dts = np.diff(onset)*dt

# Calculate the variability, over nn heartbeats
sds = []
for ii in range(len(dts)-nn):
    sds.append(np.std(dts[ii:ii+nn]))

# Print minimum and maximum variability
print(f'Minimum HRV: {np.min(sds):6.3f}')
print(f'Maximum HRV: {np.max(sds):6.3f}')
```

# B.7 Solutions to Statistics

## 1. Exercises 1-4

Listing B.19: S7_simple_stats.py

```python
""" Solution Exercises Chapter 'Statistics' """

# author:   Thomas Haslwanter
# date:     June-2020

# Import the required packages
import numpy as np
from numpy import random
import matplotlib.pyplot as plt
from scipy import stats
from typing import Tuple


def generate_and_analyze() -> Tuple:
    """Task1: Generate and analyze the data, and return them"""

    # Generate the data ...
    np.random.seed(12345)   # Ensure reproducability
    num_data = 100
    group1, group2, group2_after = {},{},{}
    group1['data'] = 8 * random.randn(num_data) + 60
    group2['data'] = 10* random.randn(num_data) + 55
    group2_after['data'] = group2['data'] + 0.25 + 1*random.randn(num_data)


    # ... and analyze them
    group1_10 = {'data':group1['data'][:10],
                 'mean':np.mean(group1['data'][:10]),
                  'std': np.std(group1['data'][:10], ddof=1),
                  'sem': stats.sem(group1['data'][:10]) }

    group1 = {'data':group1['data'],
              'mean':np.mean(group1['data']),
               'std': np.std(group1['data'], ddof=1),
               'sem': stats.sem(group1['data']) }

    group2 = {'data':group2['data'],
              'mean':np.mean(group2['data']),
               'std': np.std(group2['data'], ddof=1),
               'sem': stats.sem(group2['data']) }

    return (group1, group1_10, group2, group2_after)


def show(data: Tuple) -> None:
    """Show errorplots with SD and SEM, and a boxplot of the data"""

    # Unravel the data
    (group1, group1_10, group2, group2_after) = data

    fig, axs = plt.subplots(1,3)

    # The first two plots show SDs and SEMs ...
    ylims = [30, 90]
    for ii, param in enumerate(['std', 'sem']):
```

```python
        axs[ii].errorbar([1,2], [group1['mean'], group2['mean']], yerr=[group1[
            param], group2[param]], fmt='o g')
        axs[ii].set_xlim([0.5, 2.5])
        axs[ii].set_ylim(ylims)
        axs[ii].set_xticks([1,2])
        axs[ii].set_xticklabels(['Group1', 'Group2'])
        axs[ii].set_title('Errorbars: '+param)

    axs[0].set_ylabel('Weight [kg]')
    axs[1].set_yticklabels([])

    # ... and then the boxplots
    data = [group1['data'], group2['data']]
    axs[2].boxplot(data)
    axs[2].set_xticks([1,2])
    axs[2].set_xticklabels(['Group1', 'Group2'])
    axs[2].set_ylim(ylims)
    axs[2].set_yticklabels([])
    axs[2].set_title('Boxplot')

    plt.show()


def compare(data: Tuple) -> None:
    """ Compare groups """

    # Unravel the data
    (group1, group1_10, group2, group2_after) = data

    _,p = stats.ttest_ind(group1['data'], group2['data'])
    if p < 0.05:
        print('There is a significant difference between group1 and group2.')
    else:
        print('No significant difference between Group1 and Group2.')

    _,p = stats.ttest_rel(group2['data'], group2_after['data'])
    if p < 0.05:
        print('The carb diet causes a significant weight change.')
    else:
        print('The carb diet causes NO significant weight change.')


if __name__ == '__main__':

    data = generate_and_analyze()
    show(data)
    compare(data)
```

## 2. Gait Analysis

Listing B.20: S7_gait.py

```python
""" Solution Exercise 'Gait Analysis', Chapter 'Statistics' """

# author:   Thomas Haslwanter
# date:     April-2021

# Import the required packages
import numpy as np
import os
```

```
import matplotlib.pyplot as plt
from scipy import io

# Set the parameters
in_dir = r'..\..\data'
file_name = 'gait.mat'

# Get the data
in_file = os.path.join(in_dir, file_name)
data = io.loadmat(in_file)

# Extract them from the Matlab format
time = data['Gaitdata']['time'][0][0].ravel()
gait = data['Gaitdata']['kneeAngle'][0][0].ravel()
heel_strike = data['Gaitdata']['heelStrike'][0][0].ravel()

# Show the original data
plt.plot(time, gait)
plt.xlabel('Time (s)')
plt.ylabel('Knee-Angle')
plt.show()

# Convert the heel-strike information from time (s) to an integer index
sample_rate = 1/(time[1] - time[0])
heel_strike_idx = np.int32(np.round(heel_strike * sample_rate))

# Make one interpolated cycle exatly 101 long, so that we can interpret it
# as percent (from 0 to 100). Note that the last point is included, since
# the heel-strikes also include the first AND last point
n_interp = 101

steps = []
for ii, step_length in enumerate(np.diff(heel_strike_idx)):
    steps.append(np.interp(np.arange(n_interp),
                    np.linspace(0, n_interp, step_length+1),
                    gait[heel_strike_idx[ii]:(heel_strike_idx[ii+1]+1)] ))

# Convert from list to array, and calculate mean and std
data = np.array(steps)
mean = np.mean(data, axis=0)
std = np.std(data, axis=0, ddof=1)

# Show the result
plt.plot(mean, label='mean')
plt.plot(mean + 2*std, ls='dashed', label='95%-CI')
plt.plot(mean - 2*std, ls='dashed', color='C1')
plt.legend()
plt.xlabel('Gait-cycle (%)')
plt.ylabel('Knee-angle (deg)')
plt.gca().margins(x=0)
out_file = 'gait.jpg'
pil_kwargs = {'quality': 90}
plt.savefig(out_file, dpi=200, pil_kwargs=pil_kwargs)
print(f'Result saved to {out_file}')
plt.show()

# the same plot, but with a shaded patch for the CIs
plt.plot(mean, label='mean')
#plt.plot(mean + 2*std, ls='dashed', label='95%-CI')
#plt.plot(mean - 2*std, ls='dashed', color='C1')
plt.fill_between(np.arange(len(mean)), mean-2*std, mean+2*std,
                    alpha=0.2, label='95%-CI')
plt.legend()
```

```python
plt.xlabel('Gait-cycle (%)')
plt.ylabel('Knee-angle (deg)')
plt.gca().margins(x=0)
out_file = 'gait.jpg'
plt.savefig(out_file, dpi=200, pil_kwargs=pil_kwargs)
print(f'Result saved to {out_file}')
plt.show()
```

# B.8    Solutions to Parameter Fitting

**Listing B.21: S8_parameter_fits**

```python
""" Solution Exercises Chapter 'Parameter Fits' """

# author:   Thomas Haslwanter
# date:     April-2021

# Import the standard packages
import numpy as np
import matplotlib.pyplot as plt
import pandas as pd
import os

# For the fitting
import statsmodels.formula.api as smf


def get_data(in_file: str=None) -> pd.DataFrame:
    """Get data from NOAA about the global CO2-levels

    Parameters
    ----------
        in_file : Name of locally stored data-file. If 'in_file' is 'None',
            the data are retrieved from the web

    Return
    ------
        df : in_data, with the column names
            ['Year', 'index', 'date', 'avg', 'co2', 'trend', 'nr_days']
    """

    if in_file is None:
        # You can also easily work with the latest data from the web:
        print('Getting the data from the web')

        ftp_address = 'aftp.cmdl.noaa.gov'
        remote_dir = 'products/trends/co2'
        remote_file = 'co2_mm_mlo.txt'
        local_file = 'co2.txt'

        ftp = FTP(ftp_address)
        ftp.login(user='', passwd='')
        ftp.cwd(remote_dir)

        lf = open(local_file, 'wb')
        ftp.retrbinary('RETR ' + remote_file, lf.write, 1024)
        lf.close()
        ftp.quit()

        print(f'Data saved locally, as {local_file}')
```

```
        else:
            if os.path.exists(in_file):
                local_file = in_file
                print('Using local data ')
            else:
                raise IOError(f'{in_file} does not exist!')

        df = pd.read_csv(local_file, header=None, skiprows=72,
                delim_whitespace=True)
        df.columns = ['Year', 'index', 'date', 'avg', 'co2', 'trend', 'nr_days']

        return df


def polynomial_fits(data: pd.DataFrame):
    """Linear, quadratic and cubic fits to the data

    Parameters
    ----------
        data : input data, from 'get_data'
    """

    p_1 = np.polyfit(data.date, data.co2, 1)
    p_2 = np.polyfit(data.date, data.co2, 2)


    # Since there are numerical problems with large-x-values,
    # center them around 2000
    data['year2000'] = data['date']-2000
    explanation = """Fitting a polynomial with a large offset on the x-axis can
        lead to numerical instabilities.
    To avoid that problem, we subtract the main bias of 2000 years.
    This is part of the process of "normalization", which is commonly used in
        areas such as machine learning
    to optimize the numerical results. """
    print(explanation)
    p_2_year2000 = np.polyfit(data.year2000, data.co2, 2)
    p_3_year2000 = np.polyfit(data.year2000, data.co2, 3)


    # Show the quadratic fit-values
    print(f'\nquadratic fit: {p_2}')
    print(f'quadratic fit around 2000: {p_2_year2000}\n')

    # Fitted polynomials
    fit_x = np.linspace(np.min(data.date), np.max(data.date), 100)
    fit_x_year2000 = fit_x - 2000

    fit_y_1 = np.polyval(p_1, fit_x)
    fit_y_2 = np.polyval(p_2_year2000, fit_x_year2000)
    fit_y_3 = np.polyval(p_3_year2000, fit_x_year2000)

    # Show the data and the fits
    plt.plot('date', 'co2', data=data, label='measurements')
    plt.plot(fit_x, fit_y_1, label='linear fit')
    plt.plot(fit_x, fit_y_2, label='quadratic fit')
    plt.plot(fit_x, fit_y_3, label='cubic fit')

    plt.legend()
    plt.show()


def CIs_and_residuals(data: pd.DataFrame) -> dict:
```

```python
"""Use 'statsmodels' to find confidence-intervals, and plot the residuals

Parameters
----------
    data : input data, from 'get_data'

Returns
-------
    residuals : x/y-values for the residuals
"""


# Linear fit
mod = smf.ols(formula='co2 ~ year2000', data=data)
res_1 = mod.fit()
# print(res_1.summary())

# If you only want the confidence intervals, you get them with
ci = res_1.conf_int()
ci.columns = ['Lower', 'Upper']
print(f'The CIs for the linear fit are {ci}')

# Quadratic fit
mod = smf.ols(formula='co2 ~ year2000 + I(year2000**2)', data=data)
res_2 = mod.fit()
print('\nQuadratic fit ------------------------\n')
print(res_2.summary())

# Cubic fit
mod = smf.ols(formula='co2 ~ year2000 + I(year2000**2) + I(year2000**3)',
        data=data)
res_3 = mod.fit()
#print(res_3.summary())

# Which ones are significant?
print('\Which order of fit do we need?')
for (res, order) in zip([res_1, res_2, res_3],
    ['linear', 'quadratic', 'cubic']):
    ci = res.conf_int()
    ci.columns = ['Lower', 'Upper']
    if ci.iloc[-1].prod() < 0:
        print(f'\nThe {order} fit is not significant.')

# For the quadratic fit, show the residuals
plt.plot(data.year2000, res_2.resid, '.')
plt.title('Residuals for the quadratic fit')
plt.show()

# Select a range with an approximately constant offset, around 2010
good_years = (data.year2000>4) & (data.year2000<16)

residuals = {}
residuals['x'] = data.year2000[good_years]
residuals['y'] = res_2.resid[good_years]

return residuals




# Prepare the data for the sine-fit
phi = np.deg2rad(np.arange(len(sim_x))*30)
data_sine = pd.DataFrame({'phi':phi, 'sine':np.sin(phi),
'cosine':np.cos(phi), 'resid':sim_y})
```

```python
        # Make the sine-fit
        mod_sine = smf.ols(formula='resid ~ sine + cosine', data=data_sine)
        res_sine = mod_sine.fit()

        fit = res_sine.params
        amp = np.sqrt(fit.sine**2 + fit.cosine**2)
        delta = np.arctan2(fit.sine, fit.cosine)

        print(f'\nAmplitude of annual CO2-variations: {amp:5.3f}')


def sinefit(data: pd.DataFrame):
    """Make a sine-fit

    Parameters
    ----------
        data : residuals, from 'CIs_and_residuals'
    """

    # Prepare the data for the sine-fit
    phi = np.deg2rad(np.arange(len(data['x']))*30)
    data_sine = pd.DataFrame({
        'phi':phi,
        'sine':np.sin(phi),
        'cosine':np.cos(phi),
        'resid':data['y'] })

    # Make the sine-fit
    mod_sine = smf.ols(formula='resid ~ sine + cosine', data=data_sine)
    res_sine = mod_sine.fit()

    fit = res_sine.params
    amp = np.sqrt(fit.sine**2 + fit.cosine**2)
    delta = np.arctan2(fit.cosine, fit.sine)
    offset = fit.Intercept

    # Show values and fit
    plt.plot(phi, data['y'], '.-', label='values')
    plt.plot(phi, offset + amp * np.sin(phi + delta), label='fit')
    plt.legend()
    plt.show()

    print(f'\nAmplitude of annual CO2-variations: {amp:5.3f}')


if __name__ == '__main__':
    data_dir = '../../data'
    file_name = 'co2_mm_mlo.txt'
    in_file = os.path.join(data_dir, file_name)

    data = get_data(in_file)
    polynomial_fits(data)
    res_val = CIs_and_residuals(data)
    sinefit(res_val)
```

# B.9   Solutions to Spectral Signal Analysis

## 1. Power Spectrum

**Listing B.22: S9_power**

```python
""" Solution Exercise 'Power Spectrum', Chapter 'Spectral Analysis' """

# author:   Thomas Haslwanter
# date:     April-2021

# Import the required packages
import numpy as np
from numpy import random
import matplotlib.pyplot as plt
from scipy import stats

# Generate the original data
t = np.arange(0, 10, 0.1)
x = np.sin(t)
fft = np.fft.fft(x)
amps = np.abs(fft)

# - The FFT has many frequency components, since the end of the sine-wave
#   does not match the beginning of the sine-wave. This is some form of
#   "spectral leakage"
# - In order to get an FFT with only one component, I have to type e.g.

# Generate the signal
duration = 2*np.pi
t = np.linspace(0, duration, 100)
x = np.sin(t)

# Calculate the power of the signal
fft_ideal = np.fft.fft(x)
amps_ideal = np.abs(fft_ideal)

# Show the results
plt.plot(amps_ideal**2, '*-')
plt.show()

# - The first value is real, since it is proportional to the offset.
# - The second and third values are the amplitude and phase of the first two
#   frequency components.
# - The frequency units are:
df = 1/duration

print('\nSolution Exercise 1 ------------------------');
print(f'The first three components are: {fft[:3]}')
print(f'The offset is: {amps[0]/len(x)}')
print('The first two oscillating components have the amplitudes: ' +
        f'{amps[1:3]}')
print('The phases of the first two components are: ' +
        f'{np.angle(fft[1:3])} (rad)')
print(f'The first two frequencies are: {[df, 2*df]} (Hz)')
```

## 2. Hand-coded Fourier Transform

**Listing B.23: S9_fft_handcoded**

```python
""" Solution Exercise 'Handcoding the FFT', Chapter 'Spectral Analysis' """

# author:   Thomas Haslwanter
# date:     June-2020

# Import the required packages
import numpy as np
from numpy import random
import matplotlib.pyplot as plt
from scipy import stats

# Enter the data
dt = 0.1
N = 100
t = np.arange(N)*dt
x = np.sin(t) + 3*np.cos(3*t)

# Calculate the DFT by hand
F = []
for ii in range(len(t)):
    F.append(0)
    for jj in range(len(t)):
        F[-1] += x[jj] * np.exp(-2*np.pi*1j*(ii)*(jj)/len(t))
F = np.array(F)
# Equivalent, elegant way to calculate the DFT:

n = np.arange(N)
k = np.arange(N)
n = np.atleast_2d(n)
k = np.atleast_2d(k)
dft = x @ np.exp(-2j*np.pi/N)**(n.T@k)

# And the calculation with the fft-function:
myFFT = np.fft.fft(x)

# Show the first three values

print('\nSolution Exercise 2 ------------------------');
print('First 3 FFT components, calculated by hand (2 methods):');
print(F[:3])
print(dft[:3])

print('First 3 FFT components, calculated with the function "fft":');
print(myFFT[:3])
```

## 3. Your Voice

**Listing B.24: S9_voice.py**

```python
""" Solution Exercise 'Your voice', Chapter 'Spectral Analysis'

This example should be done with your own voice.
To show how to proceed, I use the sound in 'vowels.wav'.
"""

# author:   Thomas Haslwanter
# date:     April-2021

# Import the required packages
```

```python
import numpy as np
from numpy import random
import matplotlib.pyplot as plt
from matplotlib import cm
from scipy import signal
import os
import sksound

# Set the parameters
in_dir = r'..\..\data'
file_name = 'vowels.wav'

# Get the data
in_file = os.path.join(in_dir, file_name)

# I perfer to work with "scikit-sound". But one can also read in the data
# directly with scipy.io
sound = sksound.sounds.Sound(in_file)

# Show the spectrogram
plt.specgram(sound.data, NFFT=512, Fs=sound.rate, noverlap=256,
 cmap=cm.jet)
plt.xlabel('Frequency (Hz)')
plt.ylabel('Power spectrum ()')
plt.show()

# Select a single vowel
o_sound = sound.data[265000:285000]

# Calculate the power-spectrum
f, Pxx = signal.periodogram(o_sound, fs=sound.rate)

# Show the result
plt.semilogy(f, Pxx)

# Format the plot
plt.ylim(1e-6, 1e2)
plt.xlim(0, 1000)
plt.title('oooooooooh')
plt.xlabel('Freq (Hz)')
plt.ylabel('Power spectrum ()')

plt.show()
```

## B.10    Solutions to Solving Equations of Motion

For a spring, force and displacement are related by $f(t) = k * x(t)$. The corresponding equation for a dashpot is $f(t) = r * \dfrac{dx(t)}{dt}$.

Applying a Laplace transform, these turn into $F(s) = k * X(s)$, and $F(s) = r * s * X(s)$, respectively.

In a Voigt element, the two elements are put in parallel, and $F_{total} = F_{spring} + F_{dashpot}$. With $F$ as input and $X$ as output, the transfer function is

$$\frac{X(s)}{F(s)} = \frac{1}{k + r * s}$$

For a serial connection of spring and dashpot, the two transfer functions have to be multiplied:

$$\frac{X(s)}{F(s)} = \frac{1}{kr * s}$$

And when two Voigt elements are connected in series, the two transfer functions have to be multiplied:

$$\frac{X(s)}{F(s)} = \frac{1}{k + r * s} * \frac{1}{k + r * s} = \frac{1}{k^2 + 2\,k\,r * s + r^2 * s^2}$$

**Listing B.25: S10_mechanical_systems**

```
""" Solution Exercises Chapter 'Equations of Motion' """

# author:    Thomas Haslwanter
# date:      June-2020

# Import the required packages
import numpy as np
import matplotlib.pyplot as plt
import control
from collections import namedtuple

# Define the parameters
r = 2
k = 1

# Generate the transfer functions for spring and damper
sys_spring = control.TransferFunction([k], [1])
sys_damper = control.TransferFunction([r,0], [1])

# Combine these appropriately
voigt = 1/(control.parallel(sys_spring, sys_damper))
serial = 1/(control.series(sys_spring, sys_damper))
voigt_serial = control.series(voigt, voigt)

# Display the transfer functions
print(voigt)
print(serial)
print(voigt_serial)

# Give the output simple names, and specify the time axis
Response = namedtuple('Response', ['t', 'x'])
t_out = np.arange(0,10,0.1)
```

```python
# Simulate a step response of these three systems
r_voigt = Response(*control.step_response(voigt, t_out))
r_serial = Response(*control.step_response(serial, t_out))
r_voigt_serial = Response(*control.step_response(voigt_serial, t_out))

# Plot the solutions
plt.plot(r_voigt.t, r_voigt.x, label='Voigt')
plt.plot(r_serial.t, r_serial.x, label='Serial')
plt.plot(r_voigt_serial.t, r_voigt_serial.x, label='Voigt-Serial')

plt.legend()
plt.xlabel('Time [sec]')
plt.ylabel('System Responses')
plt.show()
```

## B.11    Solutions to Useful Programming Tools

Listing B.26: PySimpleGUI.py

```python
""" Solution to Exercise Chapter 'Programming Tools'

Demonstrates one way of embedding Matplotlib figures into a PySimpleGUI window.

Basic steps are:
 * Create a Canvas Element
 * Layout form
 * Display form
 * Draw plots onto convas

Based on information from:
https://matplotlib.org/3.1.0/gallery/user_interfaces/embedding_in_tk_sgskip.html
"""

# author:   Thomas Haslwanter
# date:     June-2020

# Import the required packages
import numpy as np
import matplotlib.pyplot as plt
from matplotlib.backends.backend_tkagg import FigureCanvasTkAgg
import PySimpleGUI as sg


def make_sine(amp, freq, duration=2, dt=0.01):
    """Generate a Matplotlib-figure with a sinewave"""

    t = np.arange(0, duration, dt)
    omega = 2 * np.pi * freq
    x = amp * np.sin(omega * t)
    return t, x


def draw_figure(canvas, figure, loc=(0, 0)):
    """ Matplotlib helper code """

    graph = FigureCanvasTkAgg(figure, master=canvas)
    graph.draw()
    listbox = graph.get_tk_widget()
    listbox.pack(side='top', fill='both', expand=1)
    return graph

if __name__ == '__main__':
    # Make the first plot
    t, x = make_sine(amp=1, freq=1)
    fig, ax = plt.subplots(1,1)
    ax.plot(t,x)

    ax.set_xlabel('Time [sec]')
    ax.set_ylabel('Sine-wave')
    figure_x, figure_y, figure_w, figure_h = fig.bbox.bounds

    # Without the following line, I get on my Linux system the ominous error
    # _tkinter.TclError: bad screen distance "640.0"
    # Since I have found no recent references to that error on  Google,
    # I just override the setting here
    (figure_w, figure_h) = (600, 400)
```

```
# define the window layout
layout = [[sg.Text('GUI-demo', font='Any 18')],
          [sg.Canvas(size=(figure_w, figure_h), key='canvas')],
          [sg.Text('Amplitude '), sg.InputText('1')],
          [sg.Text('Frequency'), sg.InputText('1')],
          [sg.Button('Ok'), sg.Button('Cancel')] ]

layout = [[sg.Text('GUI-demo', font='Any 18')],
          [sg.Canvas(size=(figure_w, figure_h), key='canvas')],
          [sg.Text('Amplitude '), sg.InputText('1')],
          [sg.Text('Frequency'), sg.InputText('1')],
          [sg.Button('Ok'), sg.Button('Cancel')] ]

# create the form and show it without the plot
window = sg.Window('Demo Application - Embedding Matplotlib In PySimpleGUI',
                   layout, finalize=True)

# add the plot to the window
graph = draw_figure(window['canvas'].TKCanvas, fig)

event, values = window.read()
while event == 'Ok':
    # Update the figure with the new parameters, until
    # the user hits "Cancel"
    amp = np.float(values[0])
    freq = np.float(values[1])
    t_new, x_new = make_sine(amp, freq)

    ax.cla()
    ax.plot(t_new, x_new)
    graph.draw()

    event, values = window.read()
else:
    window.close()
```

# Appendix C

# Abbreviations

**1D** one-dimensional

**2D** two-dimensional

**ASCII** American Standard Code for Information Interchange

**ARMA** AutoRegressive Moving Average (estimation method for the spectral density)

**BMI** Body Mass Index

**CDF** Cumulative Distribution Function

**CI** Confidence Interval

**CQ** code quantlet (source code in folder `sapy/src/code_quantlets`)

**dB** deciBel (logarithmic gain factor)

**DFT** Discrete Fourier Transform

**DSP** Digital Signal Processing

**ECG** Electrocardiogram

**EMG** Electromyogram

**FFT** Fast Fourier Transform

**FIR** Finite Impulse Response

**GUI** Graphical User Interface

**HTML** Hypertext Markup Language

**Hz** Hertz (unit for frequency)

**IDE** Integrated Development Environment

**IIR** Infinite Impulse Response

**IMU** Inertial Measurement Unit

**IO** Input/Output

**IQR** Inter-Quartile-Range

**JPG/JPEG** Joint Photographic Experts Group

**KDE** Kernel Density Estimation

**ML** Machine Learning

**MS** Microsoft

**LOESS** LOcal regrESSion

**LOWESS** Locally Weighted Scatterplot Smoothing

**NAN** Not A Number

**PDF** Portable Document Format or Probability Density Function (it should be clear from the context which of the two meanings applies)

**PNG** Portable Network Graphics

**PPF** Percentile-Point Function

**RGB** Red-Green-Blue (stack order in color images)

**RMS** Root Mean Squared

**SQL** Structured Query Language

**SD** Standard Deviation

**SE** Structural Element

**SEM** Standard Error of the Mean

**SS** Sum of Squares

**STFT** Short Time Fourier Transform

**SVG** Scalable Vector Graphic

**SVR** Support Vector Regression

**TIFF** Tagged Image File Format

**WAV** WAVeform audio file format

# Appendix D

# Web Ressources

**anaconda** https://www.anaconda.com/products/individual

**astroem and murray: feedback systcms** http://www.cds.caltech.edu/~murray/amwiki
Free online book on control systems

**bokeh** https://bokeh.org/

**control** https://python-control.readthedocs.io/

**ffmpeg** http://ffmpeg.org

**git** https://git-scm.com/

**github** https://github.com/

**gohlke** http://www.lfd.uci.edu/~gohlke/pythonlibs/

**ipython** http://ipython.org/

**julius o smith III: DSP-books** https://ccrma.stanford.edu/~jos/ Free online books on
basics and applications of FFTs

**jupyter** https://jupyter.org/

**matplotlib** https://matplotlib.org/

**numpy** https://numpy.org/devdocs/

**OpenCV** https://opencv.org/

**pandas** https://pandas.pydata.org/docs/

**plotly** https://plot.ly/

**pycharm** https://www.jetbrains.com/pycharm/

**pypi** https://pypi.org/

**pyqtgraph** http://pyqtgraph.org/

**pysimplegui** https://github.com/PySimpleGUI/PySimpleGUI

**pytest** https://docs.pytest.org

**python** https://www.python.org/

**python tutorial** https://docs.python.org/3/tutorial/

**python style guide** https://pep8.org/

**real python** https://realpython.com/

**regular expresseions - cheatsheet** https://www.debuggex.com/cheatsheet/regex/python

**regular expressions** http://www.regular-expressions.info/

**scikit-image** https://scikit-image.org/

**scikit-learn** https://scikit-learn.org/

**scikit-sound** http://work.thaslwanter.at/sksound/html/

**scipy** https://www.scipy.org/

**scipy lecture notes** https://scipy-lectures.org/

**spyder** https://www.spyder-ide.org/

**stackoverflow** https://stackoverflow.com/

**sapy** https://github.com/thomas-haslwanter/sapy

**tortoisegit** https://github.com/TortoiseGit/TortoiseGit

**visual studio code** https://code.visualstudio.com/

**wing** http://www.wingware.com/

**winpython** https://winpython.github.io/

# Index

Printed in the United States
by Baker & Taylor Publisher Services